SUSTAINABLE SOLUTIONS

SUSTAINABLE SOLUTIONS

The Climate Crisis and the Psychology of Social Action

ROBERT G. JONES, PHD

 AMERICAN PSYCHOLOGICAL ASSOCIATION

Copyright © 2022 by the American Psychological Association. All rights reserved. Except as permitted under the United States Copyright Act of 1976, no part of this publication may be reproduced or distributed in any form or by any means, including, but not limited to, the process of scanning and digitization, or stored in a database or retrieval system, without the prior written permission of the publisher.

The opinions and statements published are the responsibility of the author, and such opinions and statements do not necessarily represent the policies of the American Psychological Association.

Published by
American Psychological Association
750 First Street, NE
Washington, DC 20002
https://www.apa.org

Order Department
https://www.apa.org/pubs/books
order@apa.org

In the U.K., Europe, Africa, and the Middle East, copies may be ordered from Eurospan
https://www.eurospanbookstore.com/apa
info@eurospangroup.com

Typeset in Minion by Circle Graphics, Inc., Reisterstown, MD
Printer: Gasch Printing, Odenton, MD

Cover Designer: Anne C. Kerns, Anne Likes Red, Inc., Silver Spring, MD

Library of Congress Cataloging-in-Publication Data

Names: Jones, Robert G. (Robert Gordon), 1955- author.
Title: Sustainable solutions : the climate crisis and the psychology of
  social action / by Robert G. Jones.
Description: Washington, DC : American Psychological Association, [2022] |
  Includes bibliographical references and index.
Identifiers: LCCN 2022003019 (print) | LCCN 2022003020 (ebook) |
  ISBN 9781433834462 (paperback) | ISBN 9781433834479 (ebook)
Subjects: LCSH: Climatic changes--Psychological aspects. | Social
  action--Psychological aspects. | Social psychology. | BISAC: PSYCHOLOGY /
  Social Psychology | SCIENCE / Environmental Science (see also
  Chemistry / Environmental)
Classification: LCC BF353.5.C55 J65 2022 (print) | LCC BF353.5.C55
  (ebook) | DDC 155.9/15--dc23/eng/20220317
LC record available at https://lccn.loc.gov/2022003019
LC ebook record available at https://lccn.loc.gov/2022003020

https://doi.org/10.1037/0000296-000

Printed in the United States of America

10 9 8 7 6 5 4 3 2 1

Contents

1. An Applied Psychology Approach to Managing Ourselves — 3
2. Beyond Deep Adaptation — 23
3. Human Exceptionalism — 43
4. Psychological Factors Required for Social Speciation — 57
5. Social Margins, Conflict, and Developmental Change — 75
6. Making and Managing Social Ecotones — 99
7. Using Social Speciation on Purpose — 117
8. Defining Successful Social Speciation — 145
9. Taking Action — 169
10. Solutions — 183

References — 193
Index — 221
About the Author — 235

SUSTAINABLE SOLUTIONS

1

An Applied Psychology Approach to Managing Ourselves

I share the dire sense of urgency that many of you feel about environmental destruction. I wrote this book because we are frustrated with the lack of meaningful progress in dealing with the waste and depletion of essential resources (e.g., air, water, arable land) that characterize human activities and the inequities of the consequences of these problems of sustainability.

Unlike most, I approach these issues from the direction that Rachel Carson called for in 1962: We need to "manage ourselves." Human decision making and behavior are the causes of our problems, so we need to address these first. While evolutionary biology, history, anthropology, and a few other areas focus on describing the world in human hands, applied psychology provides more direct guidance on how to create new social species. Other applied disciplines—medicine, political science, and urban planning—all rely on sciences: evolutionary biology, history, and geography, respectively. And, most certainly, knowledge from these areas has been

https://doi.org/10.1037/0000296-001
Sustainable Solutions: The Climate Crisis and the Psychology of Social Action, by R. G. Jones
Copyright © 2022 by the American Psychological Association. All rights reserved.

used to support human evolutionary hacking. Knowledge that can be found only in these disciplines will make hack-management possible. But only knowledge and methods from applied psychology are readily available at this time to support efforts to directly manage decision making with an extensive research base. None of these other fields currently provide evidence that is so essential to this endeavor.

As an industrial and organizational (I/O) psychologist, I have been researching decision making in organizations for more than 30 years in both academic and workplace settings. I have also served on the National League of Cities Steering Committee for Energy, Environment, and Natural Resources, so I have had to squarely face sustainability problems in a policy-making role. Instead of prioritizing new discoveries in climatology, ecological biology, hydrology, and other sciences that have been brought to bear so far, I bring applied psychology to this task. And why not? If we are trying to manage collective human thinking and behavior, let's use the best science we have for managing collective human thinking and behavior.

The empirical, process-based approach applied psychology takes to managing collective human behavior can play an essential role in preventing the worst effects of the environmental changes that are currently unfolding. Only a relatively small number of applied social scientists have worked on this important sustainability mission, with too little accomplished. There has been considerable basic research into the psychology of proenvironmental behavior (Steg & Vlek, 2009), mostly trying to promote behaviors that reduce resource use. I have detailed the problems with these (mostly self-report and behaviorist) studies (Jones, 2015, 2017, 2020), where the ideas for improving things are narrowly defined by researchers themselves and where much of the research is done in controlled settings with student respondents. Further, almost none of the research in sustainability and environmental education has taken the systems approach commonly used in applied research, and only a few practicing applied psychologists have published in this area (Gifford, 2011; Klein & Huffman, 2013; Ones & Dilchert, 2013). Preparing for the very different future faced by our clients, professions, and societies requires that more of us take up the call to action: in applied settings, with clients trying to improve their situations, using methods already widely used for identifying and working toward valued outcomes.

Applied psychology provides a framework for action. The scientist–practitioner model, an empirically grounded, ethically driven, client-centered approach, has earned applied psychology a place in many organizational, health care, legal, and individual decision processes (Lowman, 2013). This method starts with the seemingly simple step of talking with clients about the outcomes they are trying to manage rather than by defining these outcomes by ourselves. This applied, "criterion-centric" approach (Bartram, 2005) immerses scientist–practitioners in the decisions that follow from trying to carefully define the desired outcomes ("criteria") of stakeholders. This process defines and guides further efforts. While "sustainability" is the essential criterion in this book, realistically, it is only accomplished by addressing the many, often conflicting needs, interests, and constraints of clients and other stakeholders—their criteria.

Even the definition of sustainability demonstrates the complexity of this criterion problem. *Sustainability* is commonly defined as both species survival and well-being in this survival (Oskamp, 2000). According to this definition, our sustainability problem is immediately more than one problem. A simple example illustrates this. The explosive population growth in the "New South" in the United States coincided with the advent of air conditioning (AC) and a sprawling roadway system. AC can make individual life easier in hot southern summers, but the increased burning of fossil fuels required for this enhancement of individual well-being has contributed to climate warming. This is not good for the well-being of people in warmer countries that have little or no AC, nor is it likely to be healthy for future humans. This enhancement of individual comfort at the expense of both other individuals' well-being and future species survival illustrates a criterion dilemma. Trying to serve one set of criteria (comfort and growth of one group) creates trade-offs with other criteria (health and safety of another group). If we add to this the conflict between the criteria of people trying to live and work in hot Southern climates with the criteria of people concerned for climate change (often the same people), we start to see the tremendous complexity of the criterion dilemma.

RELYING ON PROCESSES

Applied psychology approaches the criterion dilemma mainly by focusing on processes that help identify and resolve problems as they are defined by clients in their own settings. Various models are used (e.g., the systems perspective; Katz & Kahn, 1978), but processes are the core of most applied psychological interventions (Jones, 2020), both for understanding client issues and for methodological choices. Most of these start with an assessment of clients' needs and situational constraints. Cognitive behavior therapy (in clinical psychology), product design (in ergonomics), job analysis (in I/O psychology), survey feedback (in organizational development), and other methods then engage clients and client groups in managing their decision-making processes. In fact, applied psychologists are, in essence, decision-making helpers. We follow processes that have been shown to help clients better manage their decision processes in many different social environments (e.g., families, workplaces, markets, communities, courts). Our role is not typically to make decisions but to provide methods, feedback, and science-based information to help clients through their own decision making.

This empiricist, process-based approach resembles the approach taken by other applied sciences, including modern engineering. Before engineers start their work, they carefully frame a process to follow. This process often starts with defining what ends their science-based practice is intended to address, then talking with parties who are likely to be affected by the engineering project ("stakeholders"). So, if they are trying to control the flooding of a river, modern engineers identify what this end would look like, integrating their plans with the interests of stakeholders affected by their decisions. Though the scientific theories associated with hydraulics, geology, and climatology are certainly part of effective engineering, the possibilities afforded by science need to be integrated with the human realities and multiple criteria that are central to an engineering enterprise. Applied science relies on scientific theories and methods but is driven by criteria, decision processes, and the constraints of context.

Applied psychology usually takes this same empiricist approach, identifying both the realities of the "territory" of social organization and the realities of stakeholder criteria and constraints. Going beyond an engineering model, applied psychology identifies and parses the human processes that engineers try to include in their initial stakeholder discussions and planning. This deeper understanding of human social organization and decision processes is the primary scientific tool of applied psychologists' work.

This book presents a pathway forward by using scientific knowledge to support the choices involved in social organizing and decision-making processes. These processes are central to our adaptive strategy, and understanding and managing them will improve our chances for a sustainable future. Like other organisms, humans have adapted to the challenges of our physical environment through the age-old adaptive processes of DNA (selection) and RNA (learning) change. But we have gone further, saving ourselves from the extremes of our physical environments by creating artifices to try to moderate the challenges these extremes cause. So, for example, we have harnessed fire, invented buildings with roofs and walls, harnessed condensation, and developed a host of other methods to manage the extremes of heat and cold that disable our efforts to survive.

Other animals have done some of these things as well, but the processes through which humans have moderated environmental pressures have a unique component. Improvised social organization is essential to creating the many and varied artifices that people have constructed (Wilson, 1998). Unlike other species, we are able to adapt by organizing ourselves into a huge variety of social structures (Folke et al., 2003; Henrich & Muthukrishna, 2021) that support everything from building a roof to launching a rocket to Mars. Many of these social organizing attempts have failed to moderate the effects of the physical world, but enough have succeeded spectacularly. An obvious example is the one I've already described: the use of heated and cooled buildings that many humans currently enjoy. No other species has been able to develop such an enormous variety of habitations, never mind controlling the temperatures inside them to such an extent. Essential to all of these is our ability to devise different social organizations to coordinate collective efforts. Innovations like heating and cooling have relied entirely on the formation and sustenance of new social organizations. If you don't see this as a new evolutionary method, I would ask you to think of other species that have organized themselves to construct permanent settlements on every continent.

THE TROUBLE WITH SOCIAL ORGANIZING AS AN ADAPTIVE PROCESS

One result of this adaptive strategy is that when our attempts to moderate the extremes of the physical environment are successful, humans no longer need to rely on DNA and RNA change to adapt—at least in the short term. For example, we don't need to pass along more hair to descendants (DNA change) or learn where to get fuel for a fire (learning is based on RNA change) to stay warm on cold days. We walk into a temperature-controlled building instead, thanks to prodigious social organizing of processing plants for raw materials, manufacture and production organizations, and building trades, to name only a few work-related organizations necessary to constructing temperature-controlled buildings. Unfortunately, processes in these supporting social organizations often lead to more difficulties for the people who don't enjoy the advantages of AC and heating—never mind the future humans who will have to adapt to toxic by-products, including climate change.

This points to a second result of the social organizing spree of the past couple of centuries. Continuing, without careful management, to use new social organizations to modify the physical world has led to the degradation of the environment on which we all rely. Social organizing is both an immediate, individual blessing and a collective, long-term curse. I argue in this book that meeting this sustainability challenge no longer relies so much on the innovations that new social organizations support but on managing the processes themselves through which we organize socially. This is what Rachel Carson (1962) meant by "managing ourselves," and it is an enterprise for which applied social scientists are uniquely well prepared. It is also what a few leaders in the psychology of sustainability have described in workplaces (Klein & Huffman, 2013; Ones & Dilchert, 2013), legal processes (Castro, 2012), at the community level (McKenzie-Mohr, 2000), and at the societal level (Gifford et al., 2009; Swim, Clayton,

& Howard, 2011; Swim, Stern, et al., 2011), without much notice among other applied psychologists.

This book describes the social organizing process and the conditions under which creative social organizing occurs and provides research-based ideas for managing this process in the context of sustainability. I/O psychology and its related disciplines have been working to understand and manage decision making in a huge variety of social organizations for several decades now (Rogelberg & Gill, 2004). We have developed successful, widely used methods that currently support decisions about everything from broad, initial investment choices to very particular choices about how to organize specific components of work to make it safer and more rewarding. Perhaps because these science-based practices are quite recent, few of the resources and insights at our disposal have been applied to problems of sustainability. Finding ways to apply what we have discovered about social organizing holds promise for better managing our social-adaptive processes. This book focuses on how applied social scientists—especially psychologists—can help organize approaches to addressing this adaptive problem.

TURNING THE TABLES

Humans have learned to actively adapt the world to our needs. Rather than relying only on reactive adaptations (DNA and RNA change) for survival, we have deliberately, actively altered physical and social environments to moderate their effects on these reactive adaptive processes. These alterations have given us longer, more comfortable individual lives, but they are not turning out to be very sustainable.

How can knowledge of the social organizing processes at the base of our adaptive approach be used to make our species' survival and well-being more secure? How do we turn the adaptive process that has caused sustainability problems into a tool to increase sustainability? An even broader question, which demonstrates the paradoxical nature of this problem, is whether we can use the same social organizing that has led to our problems as a way to work our way out of these problems.

Beware: This book unapologetically argues for manipulation and control of human thinking and social behavior to save our future from our current selves. As an applied psychologist, I acknowledge and subscribe to a strong professional code of ethics that limits the type and degree of manipulation psychologists should exert. In fact, in the ethically founded processes required by the scientist–practitioner model (Lowman, 2013), manipulation and control are accomplished mostly by clients themselves. A common example is the use of a survey process that provides feedback for managing corporate decisions. Ultimately, people make their own decisions—our methods help gather data from stakeholders to help these decision makers understand the realistic opportunities and constraints of their situations, as well as some of the various outcomes predicted by the data.

But, because of these same ethics, I try to weigh the interests of future stakeholders who I hope will be able to live sustainable lives on this planet. This gives great urgency to the need to take action in ways that may step beyond current ethical boundaries of what some see as "acceptable." I have argued elsewhere (Jones, 2020) that the practice in organizational development of engaging all relevant stakeholders can help to define the limits of manipulation and control. Bringing to light future stakeholders whose interests are not heard or not addressed in everyday decisions can further delimit problems of potential overstep.

I offer scientist–practitioners and their clients an approach that I believe will provide an empirically sound, ethical, and practical approach to actions that will moderate our planetary footprint—and reduce the chances of further catastrophes. Not surprisingly, this approach starts with analysis of process—in this instance, of adaptation processes, borrowing from the work of environmental biologists (Diamond, 1997; Harari, 2014; Henrich, 2015; Roberts, 2017). This analysis also draws partly from the growing body of basic research in both evolutionary psychology and environmental psychology but introduces the applied concept of social speciation. Because social organizing processes play such a key role in the adaptive strategy of our species (Henrich & Muthukrishna, 2021), I argue in Chapters 3 to 5 that social organization can be treated as a new form of speciation. This opens insights into how these same social organizing processes have already been manipulated to create a consumer economy, increase sprawl, and accelerate ecosystem collapse (Jones, 2020). In fact, applied psychological approaches have been used extensively in social organizing—and in creating sustainability problems—as well as in alleviating a host of human problems. So, much of the "manipulation" following from this idea is already being used every day on—and by—you and me. It's time to deliberately put it to use for our future planet mates!

HOW WILL WE SURVIVE?

There is no credible doubt that human collective behaviors are the causes of problems of sustainability (Swim, Clayton, & Howard, 2011). What is different about a criterion-centric approach is that, after acknowledging that there are behaviors that need to be changed, applied psychologists focus on the outcomes that we are attempting to reach before addressing these behaviors themselves. So, the first big idea in this book is that we need to focus people's efforts on the outcomes they deeply value.

This runs against the powerful human inclination to dash to causal explanations and assign blame for problems (Kahneman, 2011) without defining the problems thoroughly first. It's easy to see this when people are asked to describe "sustainability." Almost always, their descriptions will include not just an important environmental outcome (e.g., climate warming, water shortage, species extinction) but also a behaviorally based cause that needs to be addressed (e.g., fossil fuel use, water waste, habitat destruction). It's not enough to state the problem; people are compelled to leap to behavioral causes—and sometimes solutions, too—without first defining the ends we are striving toward (reduced fossil fuel use, water conservation, habitat preservation). Stating these "desired outcomes" (criteria) is the essential starting place applied psychological change processes (Jones, 2020).

"Okay," some might say, "but climate change is caused by fossil fuel use, water waste leads to fatal shortages, and habitat destruction does lead to extinction." And they'd be right. The point is that science has still failed to describe the many behavioral causes of people using lots of fossil fuels (e.g., buying gas guzzlers, running air conditioners in empty houses, building roads to fuel suburban lifestyles), wasting water (e.g., watering lawns on rainy days, allowing extraction of water without limits), and destroying habitats (e.g., not regulating mining and forestry, plowing up land to plant food crops, grooming lawns with pesticides).

This book approaches this lack of definition for the behaviors underlying sustainability problems in two ways. First, I use psychology to frame some common outcomes that reduce the chances for sustainable futures. This will help define the behavioral outcomes that people would like to change in the hope of a sustainable future. Second, I explore some of the motivational forces that drive these psychological outcomes. What makes us want to sustain the places we love, the people we care for, the world of which we are a part? This relies on the truism that humans are compelled more by fear of loss than by hope for future goods (Kahneman, 2011). Understanding how to get past this fear and elicit hope for future goods will be considered, but so will "shutting down" social innovation by painting a bleak picture of the future. Both are tools for managing the social organizing processes.

First, the criterion-centric approach requires that we consider how sustainability is usually framed behaviorally—without the immediate benefit of a psychological perspective. This is part of defining the "territory" to be "engineered."

THE MANY FACES OF CATASTROPHE

Modern media often describe big problems humans are causing for the living world. There is a lot about this that sounds familiar, not just because it is repeated so often. It's also a familiar refrain from religions going back a long time ("Repent for the end is near!"), pessimists ("Murphy's Law"), and advertisers who've invented some new "problem" that their product will "solve." Some people have gotten so accustomed to hearing these warnings that they have stopped attending to them (Gifford, 2011).

Is it any wonder, then, that people have doubts about the urgency of problems called (variously) "pollution," "conservation," "ecology," "global warming," "climate change," "sustainability," and "adaptation"? There's been so much talk and so little apparent change made by those most able to change things (i.e., politicians, educators, business organizations) that it's easy to forgive people who dismiss these problems as "just another sales pitch" or "fearmongering." "If people really cared," they think, "why haven't things changed?"

Even for people who accept that there are problems of sustainability, it only makes sense to go on with our daily lives and hope that either the problems aren't as bad as they sound or that the people elected to lead us will use their positions to deal with them. We occasionally hear big sweeping solutions from these folks, such as the Green New Deal, but otherwise it seems like little, incremental changes are all we can hope for. The small changes don't seem to go far, but they could do the trick one day, right? However, maybe informed people don't hear about big, sweeping solutions because even these "little" solutions haven't worked that well. We have wind turbines, electric cars, recycling, and water protection laws. So why is the air getting worse, waste piling higher, and water getting scarcer? The real problem seems to be that none of these efforts have done enough to allay our fears.

Perhaps the reason no "big" solutions have been implemented is that the problems of sustainability that fuel people's fears are actually many, smaller, sometimes competing problems—consistent with the criterion-centric perspective. Most of the inventions humans see in their everyday lives (e.g., cell phones, TVs, cars) have made things better for individual day-to-day lives. After all, many of the people who think about these "big" issues (mostly in highly developed countries with good education systems) are living lives that the wealthiest and most powerful people a couple of hundred years ago couldn't have imagined. Many of us have reliable food in great variety, homes and entertainments that are luxurious even by 20th-century standards, long lifespans, comparatively safe jobs, spectacular (even flying) transportation, and the list goes on. Individual quality of life in wealthy countries would have been unthinkably wonderful for humans before 1900. So, what do we have to complain about? Why mess with a good thing?

The answer is, of course, that although the majority of people alive today in countries that publish books like this one have little to complain about, the people who will have something to complain about are the future humans who we will likely leave with toxic air, water, and land—considerably fewer resources to support the luxurious lifestyles some of us enjoy today. Even today, there are plenty of people who are not enjoying the "good life" and, worse, have been experiencing the effects of climate change, water shortages, and other ills resulting from the exclusion of sustainability from powerful groups' decision making.

For those of us lucky people who have thought about the problems of environmental degradation, there is a genuine concern for the rights of people other than ourselves and our "tribe." But even for those of us who do care about the life rights (not just civil rights or human rights) of future humans, there is only so much we can imagine doing about sustainability. Many who are trying to effect change have also discovered the limits of our individual impact.

Psychologists have also recognized that the "usual" approaches to change often don't take into account the scientific evidence about how we actually change. For example, most legal systems are predicated on the idea that the fear of future punishment will deter criminal behavior. Forensic psychologists have demonstrated convincingly that the possibility of future individual punishment only deters crime under some circumstances (Apel, 2013). If policy makers hope to reduce pollution (one sustainability criterion), it would help them to know that fear of punishment may only achieve this goal under certain circumstances (see Lynch et al., 2016).

This book is a challenge to think differently about sustainability, incorporating an applied psychological perspective. We have plenty of science describing the environmental problems we confront. We see the scary news everywhere, swirling in controversy. Instead of focusing on this fear and controversy, this book will get you thinking about positive change through a criterion-centric, social process approach. Unlike many self-help books, this book provides ways for you to do things to make everyone's lives better—not just yours.

A CRITERION-CENTRIC WAY FORWARD

There is a lot of good, nonscary news here. Scientific psychology provides many good answers to the big questions people are asking about sustainability. For applied scientist–practitioners, the people who are asking these big questions can be thought of as potential clients—people who are in a position to effect change but are seeking help to make decisions. Following a criterion-centric approach, it makes sense to consider the concerns ordinary people express and their attributions about the causes of sustainability problems before moving forward. Fears about sustainability often start with the latest environmental catastrophe in the making, accompanied by a variety of emotions. But these concerns are commonly followed by a pretty specific and disabling blame game. I have grouped the common behavioral attributions used to explain sustainability problems into five categories (see Table 1.1). People (a) don't pay attention to environmental issues, (b) refuse to accept that sustainability is a problem, (c) don't stop and think before acting in ways that affect sustainability, (d) won't change how they think and act to make things better, or (e) fail to act altruistically when they could. The mention of these five human causes seldom seems

Table 1.1

The Value of Five Popular Foes of Sustainability as Psychological Levers

Alleged causes of unsustainable behavior	Psychological definitions	Viability as a lever
1. Not paying attention	Perception and attention	Necessary but insufficient
2. Refusing to accept that there is a problem	Discrepancy testing Problem identification Empathy Denial	Neither necessary nor sufficient
3. Not thinking before acting	Heuristic processing Controlled processing Deliberation	Neither necessary nor sufficient
4. Refusing to learn or change	Behavior change Cognitive learning Developmental change	Necessary and occasionally sufficient
5. Failing to act	Social heuristics Conformity Authority Group processes	Necessary and sufficient

to lead to meaningful conversations about the science that addresses them (psychology), but it does help to focus on potential clients' needs.

We have good science to address these behavioral causes. Rigorous research has demonstrated how people (a) attend, (b) identify a problem, (c) think before acting, (d) change their thinking and behavior, and (e) take action (Jones, 2020). In fact, these five variables can be used, in this order, to describe the process through which people make decisions. We have large bodies of research published in reputable, peer-reviewed journals, with insightful descriptions of these components of the human decision process. Parts of this research have been applied widely in real-world settings, often profitably. That so few people are aware that this research exists is a catastrophic gap, made even more gobsmacking by the intense, persistent interest in solving these problems.

It is also problematic that this blame game follows the predictions of attribution theory. These five behavioral causes are ascribed to others: "them," not "us." Perhaps most important, though, is that the five causes are so clearly defined in terms of individual decision making—not social organizing and group decisions. Psychological science has demonstrated ways to change individual attention, problem identification, deliberative thinking, behavior, and social action, but I have argued elsewhere that the scope and practicality of these individual interventions are limited (Jones, 2020). Because our adaptive process—and the problems of sustainability that follow from it—are based on social decision making and organizing, psychology needs to refocus on these social processes if we hope to help (Swim, Stern, et al., 2011).

APPROACHING SUSTAINABILITY AS A SOCIAL, NOT AN INDIVIDUAL, PROBLEM

Evidence from I/O psychology, forensic psychology, and other applied areas strongly suggests that human adaptation relies on our social nature (Jones, 2020). We need social groups to manage the environment, for better and worse. An important example is that even though we hear about successful individual "leaders," the fact is that "leadership" is by definition a group process involving both "leaders" and "followers." In general terms, if we hope to move the dial on sustainability, we are most likely to succeed (or not) because of our social contexts. We also change our decision processes (the first four "levers" in Table 1.1) in response to our social lives. This means we need to define "taking effective action" in terms of groups of people taking action—not individuals.

This is also reflected in the practicalities of increasing sustainability using the four decision-making steps in Table 1.1. For example, a number of basic researchers have described success stories in household resource conservation (see Abrahamse et al., 2005, for a review). These researchers demonstrated the "simple" expedient of using feedback and incentives for reducing waste. For example, using water waste, this simple expedient relied on the innovations of faucet designers and manufacturers, building tradespeople learning how to install new designs, building inspection authorities approving these designs, and administrators willing to fund the demonstration studies, among others. These social organizations were not usually mentioned as part of the obvious success of these isolated, often short-lived endeavors (Abrahamse et al., 2005). Though such things as self-shutoff faucets are certainly helpful innovations, applied psychology focuses first on working with these supporting organizations.

Working through the social organizations that support innovation can have the additional advantage of developing broader support for sustainability. Providing a stakeholder voice for sustainability in decisions can create an awareness in decisions made by supporting organizations. This again draws attention to the importance of social organizations to sustainability: An individual buying a hybrid electric car may slightly reduce climate change, but working with a large corporation to convert their fleet to hybrids can have a far larger impact, based on a single party's decision.

SUSTAINABILITY IS PARTLY ABOUT TURNING "THEM" INTO "US"

In this book, I explore some of the social organizing dynamics that can change decision processes toward the broader perspective required for sustainability. I argue in Chapters 5 and 6 that the intermingling of groups to change something that matters to them—to avoid something bad or make something good happen (or both)—is the crucible for much of the innovation of the past couple of centuries. For applied psychologists to manipulate this, we need to know the psychological processes that lead to it. What are the conditions that motivate people to work through their conflicting interests to meet their common interests?

For some, the central criteria here is getting enough people to understand that people who appear different from themselves are still "us," not "them." I take a slightly different approach, asserting that the changes in social thinking are actually group changes in mental models, rather than changes just in individual behavior (conforming) or fundamental changes in social identities. The realization that people who are not yet born are really "us," not "them," is the latter—a change in identity. I argue that, while this may be a valuable realization for achieving sustainability, less is required—that changes in social behavior are predicated on shared understandings of the world, not necessarily on shared identities.

We have some good answers, but time is short, so let's get started. This book does not try to explain why so many people have missed psychological science as a way to approach our challenges. Instead, it offers a practical new perspective—now, while we may still have some time. I describe how some of the "tried and true" findings from applied psychology can (or can't) help us to adapt—and hopefully save what's left of our ecosystems in the process. I describe a way to use science-based practice to manage what Rachel Carson pointed to: the human urge to control our physical environment.

This may seem like an enormously ambitious project, but it is less so when starting with the scientist–practitioner model. While there haven't been many applications to sustainability, the last century has seen psychological science applied successfully to core parts of daily life. It requires only a glance at the "big picture" to comprehend these successes. Consider that 150 years ago, there was no science of psychology. In 2019, the National Institute of Mental Health reported that 45% of Americans sought treatment for mental illness or disorder in the past year. And this is a conservative estimate of the use of just one application—clinical services. These services are pretty efficacious in dealing with disorders that 100 years ago would have left people imprisoned, institutionalized, or worse. Granted, there are failures, but major clinical interventions are successful enough for insurance companies to cover them, alongside surgery and drugs, in the ordinary course of business.

Speaking of business, I/O psychology has grown dramatically in the past 50 years. Psychological data analytics are now widely used to help organizations make important decisions about people. I/O psychologists design and evaluate systems for hiring, training, performance management, group leadership, and a host of other consequential business decisions.

Consumer psychology has been used to get people to (a) attend to warnings, (b) recognize problems, (c) see that these problems require exercising (occasionally) some careful thought, and (d) change how they think and behave (Jones, 2015). Until now, the challenge of sustainability has been framed as trying to get people to (a) attend to scientific warnings, (b) understand that there are problems in our relationship with the planet, (c) see that these problems require exercising some careful thought before using resources, and (d) make some changes in how we think and behave with respect to waste and pollution.

I argue that, even though we have pretty clear understandings of these first four causes of unsustainable behavior, they are not as easily managed as people tend to think they are (cf. Raimi et al., 2017). But an understanding of social organizing makes it possible to address the fifth cause. Getting people to take positive social action is our most important challenge.

USING A RESEARCH BASIS FOR PRACTICE

The approach to doing this leads us into a controversy. The ethics of scientific-based practice rely on a review of the best scientific research at our disposal before addressing practical problems (Lowman, 2013). Deciding what to do about the causes of sustainability problems and how to get people to take positive action relies on a careful review of prior research. Unfortunately, the research basis for understanding creative social decision making in applied settings is in early stages (Dokko et al., 2014; Sanchez-Burks et al., 2015; Tang, 2019; Tsai et al., 2012). We don't have that much science to help guide efforts toward creative group problem solving and social action. We have just started to understand how social thinking and development have led to successful adaptation.

This leaves two options. The first is to follow the usual ethically safe route. This means starting from scratch, following the scientist–practitioner review of research, then gathering new data. This way, "next steps" can be based on data about the actual causes of sustainable social behaviors. If, however, we're in a rush to try to do something, there is a second option. We skip past the incremental data gathering and rely instead on theory. Theories are based on data from other fields and questions that look related, but we do not directly test the conditions that lead to social organizing. I take this latter path.

It is risky because it can lead to ethical lapses and also to some of the same dead ends we would have found through a more methodical approach. But there is an ethical imperative to try to catch up quickly with 70 years of work aimed at getting people to waste stuff (i.e., much of consumer psychology; see Jones, 2015), so it seems like a risk worth taking. Hopefully, we'll be able to change direction using these methods if (a) they work and (b) the ethical costs appear manageable.

Because of the urgency of environmental damage, and especially the effects of climate change on human sustainability, I step beyond the limits of existing research and venture into a theory that I believe is worth testing. To be clear, developing a theory that has some basis in existing research is still science. It is, in fact, more typical of applied science, such as consumer research. But it remains theory. If scientist–practitioners rely on it when making decisions, we are ethically constrained to do a couple of things. First, we check with independent colleagues about the ethics of the proposed approach. Second, we keep careful track of what happens as a result of our choices and report findings appropriately. I use a third precaution: keeping the ethical contingencies of this method in the forefront of discussion.

THE THEORY, IN SHORT

Here's the theory, stated as concisely as possible. I believe that there is a need for developmental change in human thinking. Such change accompanies something I call *social speciation* that is the primary adaptive advantage our species enjoys. I rely not only on the first four decision-making steps (Table 1.1)—attention, problem framing, deliberative thinking, and learning—as ways to manage the trajectory of human activity moving forward. These will still get the discussions they deserve. But I also go further, with the hope that a well-informed leap to action (the fifth human cause) is better than either a slow grind toward inevitable failure or an uninformed leap to disaster.

I believe that the same adaptive mechanisms that have gotten us to where we are today will need to be the basis for our successful future adaptation. This confronts the deep paradox already discussed. Most animals (including us) rely on internal changes in our physical structures and behaviors as ways to survive. The premise of all past evolutionary relationships has been that the structures and behaviors of a species change, partly as a reaction to the demands of their local environment and partly randomly. We adapt ourselves to meet environmental demands. These are the environmental rules that all species have been subject to, the "programming code" underlying the evolution of life—until recently.

Humans have turned things the other way around: We have learned how to adapt the world around us to meet our survival needs. We have adapted by changing the external world instead of changing our own physical structures and behaviors. Rather than changing just in the ways other animals change to meet environmental challenges—mutation, behavioral response learning, mimicry, rigid social structures, and social bonding—we have learned to conceptualize and then operate on the causes of our external challenges. In this way, we have hacked the environmental code by creating our own cozy niches. We are evolutionary hackers.

The success of these hacking operations relies on novel social structures (e.g., "teams") and social processes (especially language and science) that we devise for meeting survival challenges. Our survival mechanisms are similar to the bird nests, beehives, and prairie dog warrens other species use to meet basic needs (shelter, food, warmth). But we have used our causal reasoning and novel social interactions to deliberately alter our environment in inventive ways rather than through "rote" behaviors programmed into our genes. Thus, novel social structures and processes are our main adaptive strategy.

In a nutshell, this book describes the general theory of how this adaptive strategy happens and how it differs from other adaptive processes. This theory fits a considerable body of evidence, some of which seems obvious. It also helps to explain how we have built our current set of problems. I lay out the paradox described here in greater detail, then rely on evidence about the other four "human causes" to offer specific steps that are likely to take us toward a more sustainable future. This theory is offered to help us to readapt the unsustainable environment we have created.

SUMMARY

It is high time we address sustainability through the lens of applied psychology. Sustainability is an urgent issue confronting future humans in ways current people can only begin to fathom. Some of the psychological methods that have been successfully deployed to manage human thinking and behaviors (especially consumer psychology; Jones, 2015) have contributed to our current predicament. But this also means that we have some pretty good ideas about what will and won't work for managing sustainability. Psychological methods are just waiting to be applied to this purpose, but we need to take stock of the psychological tools that may support our efforts, both ethically and practically. Even more important, we need to define our problems using psychological outcomes as a starting point, rather than defining sustainability only in terms of outcomes that result from psychological processes. Readers will see in the chapters that follow that there is good news: People have been doing this kind of adapting throughout our time on the planet. But there is still plenty of complicated work ahead.

Beyond Deep Adaptation

Deep adaptation is a popular notion of late (Bendell, 2018). It starts with the assumption that climate change, and the cascading changes following from it, have already hit a tipping point—that there is no going back. The essential consequence of this is that humans (and other animals) will need to adapt, perhaps in ways they have never adapted before. The deep adaptation perspective argues that the "usual" means of adaptation—natural selection based on mutation, migration, learning, and physical artifice—is not going to work anymore, that we need to find new ways of adapting. Put in evolutionary terms, these inadequate forms of adaptation will lead to extinction.

I offer an alternative, psychological perspective and a paradoxical solution related to the deep adaptation idea. I argue that our success as a species is already based on a new form of adaptation, reliant on the ability to socially organize. Rather than requiring a new form of adaptation (deep

https://doi.org/10.1037/0000296-002
Sustainable Solutions: The Climate Crisis and the Psychology of Social Action, by R. G. Jones
Copyright © 2022 by the American Psychological Association. All rights reserved.

or otherwise), I argue that the path to sustainability requires a clearer understanding of our existing adaptive repertoire. In addition to the age-old adaptive mechanisms, humans have evolved a new form of adaptation, which I call *social speciation*. Social speciation is defined by the invention and construction of complex social organizations to support the construction of artifices (especially buildings and technologies) that moderate environmental pressures. In this chapter, I place social speciation within the evolutionary perspective so that it can be understood as a form of deep adaptation—that is, as a new adaptive strategy. To do this, I need to take a brief walk through an evolutionary–psychological perspective on adaptation.

First, from a biological perspective, there are some amazing ways that life has met the many challenges presented by our environments. There is a huge catalog of adaptive strategies, and, depending on the particular challenges of an ecosystem, these strategies sometimes work—species manage to survive. In fact, biology and psychology have uncovered a large number of these adaptive strategies in recent decades. For example, biologists have discovered a lot about the basis for most of these adaptations—changes in our DNA. We now know at the molecular level how the process of genetic mutation occurs. Adaptive strategies that follow from these fundamental genetic changes are quite diverse but generally boil down to one of two kinds: changes in physical form (e.g., bigger, smaller, air breathing, fur covered) and changes in behavior (e.g., seasonal migrations, instinctive building of artifice, eating meat).

These two—form and behavior—are interrelated (Heers, 2016; Lautenschlager et al., 2013). Understanding the origins of aerial flight in animals provides an example of this interplay between physical form and behavior. Flying (a set of behaviors) relies on wings (a physical form). We won't venture deeply into the complex interplay of structure and behavior that leads to such adaptation over time; this is the purview of evolutionary biology and paleontology. But I will return to the example of flight because it gives a pretty obvious illustration of how form and behavior have interacted to realize a dramatic adaptive change. So far, geneticists have successfully focused on altering physical forms (including those related to flight) and have begun to piece together some of the important ways that structural changes relate to genetic change.

But manipulating genetics to change behavior remains a tricky area, not least because of the ethical problems raised by, for example, splicing genes (Gerlai, 2001). From a purely mechanical perspective, forms do not always lead to behaviors (Heers, 2016). There is, therefore, no direct path from a change in form to a change in behavior. There is not a direct path from genes to actions. It is pretty much nonsensical to try to find the "gene for flying" or the "gene for great thinking." This is because the ways animals use their structures are determined by something else: behaviors.

BEHAVIORAL ADAPTATION

The direct study of this behavioral aspect of adaptation has led to a second, less widely recognized set of breakthroughs. This is the wide class of adaptive strategies referred to as *learning*. At the outset, it's important to understand that learning is not a single adaptive strategy or even a single set of strategies. Along with observable changes in behavior, the concept of learning is used to describe several different forms of memory, changes in motor coordination, and the development of expertise. There is, however, a commonality underlying the behavior change component of learning: Just as change in physical form relies on DNA, so behavior change appears to rely on RNA (Leighton et al., 2018). It's also helpful to understand that there is a scaffold here, where the forms developed through DNA change have led to the ability to learn. The evolution of the nervous system through DNA-based change has given rise to learning in its many forms. The relatively simple nervous systems in early animals have evolved into increasingly complex control structures within the organs referred to as *brains*.

Although several different brain structures play roles in different sorts of learning, I focus on one that is particularly relevant to human adaptation—the neocortex. As astounding as advances in genetic science have been, the past 50 years of research on learning and the human nervous system have been similarly revolutionary—if less well recognized (Gluck & Myers, 2001; Hothersall, 1983). I believe that, although we are just starting to piece together how it occurs, our best chance for survival comes from understanding how a particular aspect of human learning occurs. Starting with an understanding of learning structures, and particularly the neocortex (Iyengar, 2013) and hippocampus (Gulledge & Kawaguchi, 2007), basic research is uncovering the physiology of complex learning in multiple species (Iyengar, 2013). Just as scientists have come to understand and apply the structures that function in a coordinated fashion to produce flight (in animals and humans), I try to piece together the processes through which learning and social structures come to function to produce novel social organization. I argue that the origins of the human ability to invent novel social organizations rely on understanding how developmental change arises at the intersection of different groups' shared mental models.

THE COMMON ROOTS OF PSYCHOLOGICAL SCIENCE AND EVOLUTIONARY THEORY

It's important to see the science of psychology for what it is: a branch discipline of biology, like botany, microbiology, and zoology. By treating psychology as a branch of biology, human evolution can be seen as an ecological process in which structure and function interact in complex ways. Particularly relevant to this structure and function puzzle, psychological science reveals how essential mental and social evolution are to our current circumstances, giving a much clearer picture of the unusual processes through which we have adapted. With this knowledge in hand, we can construct a psychobiological theory that will support sustainable action.

Most important, psychology has discovered how some species adapt more quickly than others—by various kinds of learning. Genetic selection, migration, instinctive building (as in bees and ants), and other strategies studied in ecological biology occur pretty slowly. They are slow to change precisely because they rely on genetic changes that require at least one generation to occur in nature. However, learning allows for faster adaptation. Animals can sometimes learn a new behavior in a matter of minutes, far faster than the generational change required for genetic change and selection.

Think of how long it may have taken for dinosaurs to develop the physical structures necessary for flight. We don't know exactly, of course, but it was almost certainly a matter of thousands of years of mutations, false starts, and sidetracks until the behaviors of flight and the physical structures needed to fly coincided enough to send an animal into the air. Now think about how long it took from the time that people first designed the means to fly (presumably gas-filled balloons) to the point where we now routinely travel safely in jets that weigh many tons. It has taken far less time for us ill-equipped primates to fly than it did for dinosaurs, whose bodies were gradually transformed over eons into pretty well-designed flying forms.

What were the biopsychological processes required to make behavioral change happen so quickly in humans?

LEARNING IN ITS VARIOUS FORMS

As we venture into learning as a form of adaptation, psychological science is our primary source of knowledge. A large and rich body of scientific research has led to the discovery of the various kinds of learning that occur. We know, for example, that behavioral learning is related to RNA change, that it occurs even for fairly rudimentary living organisms (e.g., planaria worms, earthworms, bacteria), and that it has mechanisms that allow for "changing back" to original behaviors given the right circumstances (Bronfman et al., 2014). All of this occurs without genotypic change of any kind within a single organism during its single life. It occurs pretty much constantly throughout a human life, showing pretty definitively that you can, in fact, "teach an old dog new tricks."

Discoveries from research on behavioral learning have had powerful applications. Perhaps the best examples come from the systems engineering field. Here, engineered processes rely heavily on psychological research about perceptual systems, interconnections in the nervous system, and the feedback required for learning to occur. For example, the "learning systems" that guide self-driving cars are based on our understanding of exactly these components of human learning processes. To give an idea of the scope and power of this work, research on stimulus–response learning (originally called *behaviorism*) started with psychologists in the early 20th century and has since expanded into entirely new offshoot fields, including cognitive neuroscience, connectionism, and artificial intelligence.

In addition to elaborating how relatively simple stimulus–response learning occurs, research since then has unlocked some of the mechanisms of other forms of learning, linking these with the distinct memory systems and physical processes in our complex, primate brains. Scientists are uncovering how animals with brain structures such as the cerebellum, cerebrum, and neocortex "learn" from their experiences, a process initially called *cognitive learning*. Rather than go into great depth about how, for example, the hippocampus functions as a center for learning of "mental maps," we move directly into a couple of the characteristics that define how these brain systems have influenced human adaptation.

COGNITIVE LEARNING AND MENTAL MAPS

A quick review of the history of cognitive learning will help set the stage for understanding how the forms and functions of human adaptation have led to our social–adaptive repertoire. For a start, shared mental models (Mohammed & Dumville, 2001) serve a pivotal function in the human adaptive process. Understanding their nature may also prove crucial to our endeavor to manage sustainable action. A brief review of some of the original research that has led to the understanding of shared mental models may help in the service of this end. Apologies in advance to readers who are familiar with this literature.

Edward Tolman observed in the 1940s that some species seem to use mental maps to guide their behavior. After watching them run mazes, he posited that rats are able to construct some kind of representation of their external environment using their brains. He also posited that these mental maps explain how animals learn the cause-and-effect relationships in the world around them. In particular, he followed an earlier psychologist—Edward Thorndike—in arguing that the "law of effect" determined which cause–effect relationships were learned in a particular circumstance. The "effects" of a behavior determined what relationships got represented in mental maps. Getting cheese at the end of the maze leads a rat to connect maze running with the happy effect of cheese. This seemingly simple observation explained how a few animal species developed *contingency rules* to guide their behavior across circumstances. For example, rats learn that sniffing things that look like food before trying to eat them makes it less likely that they will get sick.

Later research confirmed that the animals that develop such causal mental maps have a thin layer of cells called the *neocortex* on the outside of their upper brains. Animals with this sensory-based learning structure are able to shortcut the long process of DNA change that leads to changes in physical form. Instead of relying on these changes in form to allow changes in behavior, animals with a neocortex can go straight to behavior change without changing specific forms to do so.

For example, some animals have developed remarkable sense organs that make them more sensitive to smells that relate to nutrition or toxic elements in their surroundings. These structures have developed through DNA changes over generations and provide an advantage to animals that have them. A favorite example of this is the turkey vulture, which can smell "food" far away, while its close cousin, the black vulture, can see objects at great distances. Between the two of them, they can find food incredibly well (Ackerman, 2020).

Animals with a neocortex have expanded their ability to integrate and elaborate from existing sensory structures. They can learn what to look for (or smell) to find their way to food—without directly smelling the food itself. They don't need quite the same specific, high-end sensory systems that help other species find food. This is a revolutionary change in form (the neocortex) that allows animals to use (behave with) their nervous systems.

While the neocortex is a remarkable structure, it is not by itself enough to serve as our adaptive secret. Contrary to some popular thought on this topic (Harari, 2014), the neocortex in itself is necessary but not sufficient for our success. But understanding this kind of learning that we and a few other species can do helps advance our understanding of our actual adaptive secret. The mental mapping that the cortex engenders, called cognitive learning, while imperfect, is an important starting point for understanding the forms and functions that lead to novel social organizing.

At the further risk of explaining things that many readers already know, let's spend a little time on this kind of learning. Most of the time, when people talk about "learning," they mean cognitive learning. In school, teachers show us words and ideas, and we try to recall them on tests. At home, we watch an elder perform some procedure, such as frying an egg, that we try to remember and replicate later (such as when we're hungry the next morning). We watch a friend getting in trouble with their parent for hurting a younger sibling and know not to hurt younger children, at least when that parent is around.

There are a few things in common across these examples of cognitive learning. First, we are learning the contingencies of behaviors. If we successfully (A) remember an idea from a class, we (B) perform well on the test. A leads to B. Cause leads to effect. This is the same for the behaviors required to cook an egg—we understand how one set of behaviors leads to the desired outcome. Several A's lead to a happy B. A string of causes leads to an effect. If we hurt another child in front of a parent, then we will be punished. A leads to B in the presence of C, with C being the presence of a parent. Cause leads to effect but only in certain circumstances. These are all examples of how some important outcome occurs as a result of behaviors in a circumstance. These causal contingencies are what we are built to learn—quickly, easily, and without much thought, we remember how behavior leads to an outcome. This is the nature of our mental maps: They are maps of the cause and effect contingencies of actions.

A second general statement about these examples may be slightly less obvious and does not hold for all cases. Much of our cognitive learning occurs through other people. In these examples, remembering what others tell us and show us and by watching what happens when others choose actions, we form cognitive maps. There is a good bit of evidence that learning of this social sort is more effective than other kinds of cognitive learning (Henrich, 2015; Jones, 2020). A more consequential example comes from medical schools, where social learning is commonly used precisely because it increases the chances that things learned in this way will be retained in practice (G. C. H. Koh et al., 2008). From here on, we will return often to the nature, processes, and consequences of this "social advantage."

COGNITIVE LEARNING FOR SUSTAINABILITY

Cognitive learning is powerful, but it is not unique to humans and clearly has not solved our sustainability problems. There are at least two reasons. First, cognitive learning is deeply enmeshed with motivations. The problem here is that we are not inclined to learn things that we do not see as meaningful or important to the situations we—and the people we are close to—are in. If self and family are affected, then we are more motivated to spend time and energy both to learn new things and to sluff off old ways of thinking and doing things. When such motivations are absent, we are inclined to heuristic and biased thinking.

But there is a less obvious motivational problem with cognitive learning when dealing with environmental problems. The prospect of profound environmental change is stressful (Reser & Swim, 2011). For some people, "emergency response" to climate catastrophes has become the new normal. Although one may be rightly concerned about the future survival of our species, responding with anxiety like this tends to narrow attention (Tay et al., 2019), focus memory on episodes rather than broader associative learning (Pergamin-Hight et al., 2015), and reduce the kinds of cognitive functioning that are most helpful for identifying novel solutions to complex problems (Lambert & McLaughlin, 2019; L. A. Thomas & LaBar, 2008). Unable to think clearly or to calmly find answers that will help cope with the situation, we fall back on well-learned automatic responses and the ability to vividly remember the stressful episode, presumably to debrief later under less stressful conditions (Marko & Riečanský, 2018; Pergamin-Hight et al., 2015).

Perhaps the most telling studies about the effects of stress on human learning and problem solving come from neurophysiological studies of the effects of stress on different memory systems. Using magnetic resonance imaging, Schwabe and Wolf (2009, 2012) showed that people under lower stress remembered information better and were less likely to revert to *procedural memory* systems. This procedural type of memory is what we use to respond quickly without thinking much—where our heuristics can be counted. The second study, by a group of psychologists at the University of California, Davis (Laredo et al., 2015), showed that behavioral flexibility was greater for male mice that had been administered morphine to relax them. Cognitive learning and trying new things both appear to be greater when stress is low.

Generally, risk aversion and heuristics will hurt more than help us in a world where there is enough to go around. Thinking back to the usual psychological foes of sustainability in Table 1.1, how does our risk-averse, heuristic-prone mind affect our (a) attention, (b) problem identification, (c) deliberative thinking, (d) cognitive learning, and (e) taking action? In fact, I have argued elsewhere that risk aversion blocks our progress with four of these five psychological culprits (Jones, 2015, 2020). We attend only to the most obvious, potentially harmful discrepancies in our immediate environment (a). We are only motivated to label something as a problem when it directly affects us and our heuristics aren't dealing with it adequately (b), and even then, we are likely to fall back on risk-averse heuristics. Otherwise, we rely entirely on heuristics, doing everything we can to *not* think deliberatively in most circumstances, trying to save our thinking power for when it's safe (c). Cognitive learning (d) occurs only under a few choice circumstances, where heuristics aren't working and when there are other people around who are trying to help us to learn.

RISK VERSUS OPPORTUNITY

Framing sustainability as an intractable, uncontrollable problem is, therefore, one of the barriers to addressing it with the sort of novel thinking that humans have used as a core part of our adaptive strategy. Automatic, stress-induced responding is highly adaptive in a threatening world, such as the environments our ancestors survived. It's important to be on the lookout for lions hunting for meat, elephants stampeding, enemy warriors looking to take over your lands, and so on (Tay et al., 2019). Safe opportunities to gather food, find a mate, or otherwise take our chances at getting what we want from the world will take a back seat to our attempts to avoid being a lion's next meal. Once we are safe, fed, and warm, we start seeking opportunities—and learning to think in novel ways. Because humans get what we want mostly as a result of cognitive learning, we only have chances to do this now and then, at least when we see the world as threatening.

This evolutionary tendency to prioritize risk over opportunity probably explains the strong evidence that our decision making is tilted toward avoiding risks even today. In one of the most important 21st-century books so far, Nobel laureate Daniel Kahneman (2011) demonstrated how our causal reasoning is fundamentally tilted toward dealing with immediate risks. With all else equal, we are inclined to make choices to "not lose"—sometimes at the expense of opportunities for gain. But is such overweighing of risk still appropriate in the relatively safe environments of today?

By framing gradual environmental degradation as a "threat" rather than an opportunity, people's attention is turned away from the opportunities at our disposal to manage ourselves. Facing people toward "opportunity framing" is more likely to lead to the sort of calm, creative responses that science—and other social learning—have yielded so far. But the inclination to focus on fear at the expense of hope reduces the likelihood of cognitive learning in the first place and inclines us toward less adaptive stress responses, given the actual urgency attached to sustainability.

PSYCHOLOGICAL DEEP ADAPTATION?

If we want deep adaptation, we need to account for these usual ways of adapting. From the perspective of evolutionary psychology, cognitive learning and heuristic processing have been essential to the adaptation of humans and other species. But, for various reasons, these strategies may not work as well for dealing with environmental degradation as they have for other challenges. First, as we've seen, there is the short circuiting of deliberative thinking and cognitive learning that results from risk aversion. Environmental destruction looks like a pretty scary challenge, which probably sends many of us into paroxysms of heuristic processing. Instead of stopping and trying to carefully work things out, people get lost in old causal explanations, heuristic processing, and remembering vivid images and well-learned procedural responses rather than trying to come up with new ways of doing things. So, cognitive learning may be more likely to contribute to our demise as the same old heuristic-building strategies have helped put us into the current situation, for better and worse.

A second reason that cognitive mapping by itself may not guide us to a sustainable future is that we've used it to build a world that forces us to rely on heuristics. In his book *Influence*, Robert Cialdini (1988) presented an early version of this conundrum. He started by pointing out that our main adaptive strategy is the ability to think things through in ways that most other species can't. We are able to come up with causal explanations and arrange our world accordingly. Rather than simply responding to its challenges through instinct and behavioral learning, we have the ability to think deliberately. For example, back in prehistory, someone observed that having some solid object above you keeps the rain off your head. This causal chain (solid object overhead→dry hair) led to a more general, abstracted idea that we can seek shelter and even construct shelters from local materials.

Cialdini (1988) argued that we have short-circuited the adaptive advantage gained by this deliberative thinking. We have built a world that makes such thinking less likely than it used to be. The rapid pace of modern life sets us back into the same kind of automatic, procedural thinking that other species use to adapt. We rely on heuristics that are the "same old approach" rather than the adaptive strategy that got us here: deliberative thought. His solution, therefore, was to get people to think more deliberatively, and he gave guidance for accomplishing this.

Adapting the World to Us

I argue that the conundrum is a bit more complicated. What we have really done to adapt to our world is to adapt it to us. Like other animals, we change individual behaviors in response to environmental conditions. But we have also changed the world to suit ourselves. We have done this by using the deliberative thinking Cialdini (1988) correctly saw as an adaptive advantage. In the dry hair example, once somebody had mentally mapped the general idea that having some solid object overhead keeps the rain off, all sorts of interesting structures were invented. Rather than ducking back into the same old dirty cave, people built roofs. Following this simplistic example, the entire field of architecture has been constructed (pardon the pun) from the "roof" idea and other ideas like it.

In her book *Tamed*, Alice Roberts (2017) described 10 species humans have domesticated over the past 20,000 years or so. These are remarkable examples of how we have gone far beyond just building roofs, wheels to drive carts, and other mechanical and technological inventions. We have effectively invented new species to reduce the many risks and uncertainties of a less domesticated world. She suggested that we have domesticated ourselves in some important ways, as well—a point I return to in the next chapter.

It is, in fact, this heavy reliance on deliberately changing the world around us to suit our needs that is the source of air pollution, waste intrusion, and shortages, among other environmental problems. We have used our ability to harness cognitive maps in our usual, risk-averse fashion. People are pretty well protected these days from lions and tigers and bears (and a host of other risks) because of this adaptive strategy. But this has also built unsustainable changes into the environments in which we live. We have become safe and comfortable as small groups, but we have not always accounted for the many impacts our approach to safety and comfort has had on the broader environmental systems on which we rely.

We've Beaten the Carrying Capacity Problem—For Now

The idea of carrying capacity is used to describe this problem. *Carrying capacity* is based on the assumption that there is a limit to the resources available for a species' survival. I suggest that this has made far less difference to humans than to other species. This is because it is based on Darwin's assumption that adaptations are specific to a particular ecological niche. But the human neocortex has provided us with an extremely broad adaptive capacity. We have used our brains to stave off selection pressures by rapidly building complex, widely applicable systems that we use to alter our local surroundings. Humans evolved in hot African savannahs but are now found in permanent settlements on every continent—a dramatic example of adapting local environments so that we avoid carrying capacity pressures.

Oxford historian Yuval Noah Harari described this differently. He argued in *Homo Deus* that we have made the world into one great big ecosystem (Harari, 2014). I suggest instead that we have used the same operations to alter every niche we want to alter. The purpose is, of course, to make ourselves safer and more comfortable wherever we want.

A look at our evolutionary story shows how the artifices we have built continue to keep us from reaching the Earth's carrying capacity. Certainly, extinctions of other, once-successful organisms show that carrying capacity still operates within ecological niches. A few species have even been able to fit harmoniously into our artificial niches. These are the birds and insects that don't "bother" us and don't provide much nutritional value (swallows, house wrens, earthworms) but are comfortable with the world we've altered. Others have changed structures and functions to join our niche adaptations—hence domesticated animals and "pests" are now "inside" our niches. But none of them can do what we have demonstrated—deliberately adapting their niches. Other species continue to try reactive adaptation (e.g., natural selection, behavioral learning, migration, mimicry), but this works only when new human niches allow this or where their old niches have not been disturbed enough to drive their residents to extinction.

Are we producing too many members of our species? Maybe so, but even if we haven't gotten there yet, some problems deserve attention now, and they may help with the carrying capacity problem down the line as well. And because our niche pretty well covers the entire planet these days (similar to Harari's point), it is sensible to expect carrying capacity problems will affect our entire niche—the Earth—at some point.

WHY DEEPLY ADAPT?

Instead of referring to our approach as "deep adaptation," I prefer the more descriptive conceptualization of humans moving from "reactive" adaptation to adaptation that relies on altering the ecological situation.

Table 2.1
Rate of Adaptive Change

Genetic mutation/ natural selection	Behavioral learning	Heuristics/ convention	Cognitive learning	Alter immediate environment	Alter genes	Manipulate altering process
−3	−2	−1	0	+1	+2	+3

Table 2.1 illustrates this in detail. Until the latest geological era—called the Anthropocene, after us—the survival equation pitted evolutionary change against the constants of an ecosystem. These constants were the finite resources and likely dangers of an ecosystem and ultimately its carrying capacity. Species altered their body structures and their behaviors in response to these challenges. Now and again, the challenges would change due to climate change, invasion of other species, and so on, but these were the "new constants" of an ecosystem. Species still had to react or die.

Now and again, a revolutionary new form with new behaviors expanded the ecosystems where life could survive. Single-cell organisms "teamed up" into mobile masses that could change location in search of food, and animalia was born. Fish developed internal respiration that gathered oxygen from the air, and amphibians formed. Reptiles grew membranes and feathers and learned to fly, and birds arrived in the air. Sometimes these revolutions led to a new dominant species; rather than finding new places to live, revolutionary forms like sauropod dinosaurs and warm-blooded mammals "took over" using the reactive evolutionary process.

All these prior, reactive revolutions still followed the basic equation, pitting structure and behavior against the constant of environmental demands. Our social speciation revolution was different. We actively, sometimes thoughtfully, and with novel social structures push back against this equation. Instead of "conforming to the rules" of our ecological situation, humans learn the rules of niches and then alter the niches to suit our wishes. The enormity of the rearrangements of niches has required novel social arrangements. Ecological demands are no longer held constant.

Instead, we open new ecological territories for ourselves by altering these environmental demands themselves. This is how we have hacked evolution.

Put into evolutionary perspective, our species has succeeded in creating a new sort of "flight." Our sociodeliberative approach is an entirely new form based on relatively new brain structures (neocortex) and processes (language and social invention). We have stumbled into a highly flexible, broadly applicable, and rapid answer to the problem of adaptation faced by all living species: how to extract enough—but not too many—resources from our environment while avoiding its inherent dangers. We have made our wishes into realities, "flying" into dreams.

A psychiatrist friend and I used to talk about something like this when we contrasted what we do professionally. Jim would talk about how sad it was that people have so much trouble adapting to their realities. Many of the problems he dealt with, relating mostly to anger, depression, and anxiety, arise from people's inability to make enough sense of their situations. Instead of making choices and taking actions that may advance their interests, his patients retreated into fantasies, leaving the realities to have their way. Because of this, they often fell into further problems that only made things worse—addiction, isolation, and so on. What he did professionally was help people to figure out more adaptive behaviors—to adjust to their circumstances.

First, I agreed with this. But I also suggested that it's pretty sad, too, that people are trying to adjust to a reality that pretty well sucks—sometimes with no happy way out. I suggested we could be framing the problem in terms of trying to adjust reality—something that industrial and organizational psychologists are trying to do in workplaces every day. It also dawned on me—much later—that this kind of reality adjustment is what humans do so well!

This realization has led me to try to answer the question of whether we can also adjust the urge to adjust reality, at least before too much "bad stuff" starts to happen. It is, of course, what you are reading about here.

This is where our paradox comes into play: Can we employ deliberate adaptation, with its highly flexible, rapid, broadly applicable systems, to avoid hitting the wall of carrying capacity? Can we use the adaptive methods that have caused our problems to solve these same problems? Answering this question will rely on understanding more about the specifics of the human structures and systems that have given rise to our revolutionary approach to adaptation. It also relies on understanding the processes through which evolutionary hacking has occurred.

Put as simply as it can be, the problem is this: We need to figure out how to use our deliberate environment-altering abilities to restructure the world that we have built. Using Jim's words, we need to "adjust to our adaptation." This does not mean going back to nature. That won't happen. We have tasted the fruits of a safer, more munificent world, and, if Alice Roberts is right, we've even changed who we are in the process. There is no going back.

But we do need to understand the workings of our environment-altering abilities if we hope to manage them better. We need to use our special kind of adaptive strategy to adapt it to our present circumstances. This means we need to know how it works, as well as how to direct and change it. We need to "manage ourselves."

RATE OF ADAPTIVE CHANGE

Thinking in terms of evolutionary adaptation, we have set the "rate of adaptive change" continuum to a positive number (see Table 2.1). At the slowest end, there is genetic change through natural selection that occurs over the course of generations (a −3 rating on speed of adaptation). In the middle, we have cognitive learning, with a 0 rating on rate of adaptation. This allows us to make broadly applicable adaptive decisions in the moment, based partly on heuristics but also on more systematic thinking. As we deliberately alter our world so that we don't have to change ourselves to meet its demands, we move into the +1 and upward rate of adaptation. As we enter the positive range, we are effectively working in the future rather than reacting to past events or responding to immediate ones. At this point in our history, having gotten around to harnessing genetic change, we are into the +2 rate of adaptive change—altering life forms before they've ever had to react to the environment. Finding our way to +3 is the challenge.

SOCIAL SPECIATION

But we have gone even further, putting our cognitive map building to use to deliberately construct the social organizations needed to build roofs, domesticate plants and animals, and change our genetic structures. Going back to the roof example, if you have ever tried to build one by yourself, you know how important social organization is. You pretty much have to have at least one other person who can coordinate with you to build this roof. And it's not like we're bees or ants that have a genetic makeup that determines whether we are the "holder of the vertical beam" or the "placers of the roof" or the "hammerers of the roof to the beam." We have to (a) figure out these social roles in the same way we figured out that roofs will keep us dry and (b) arrive at ways to influence one another to take on these roles. Put in general terms, we have used deliberative thought to organize ourselves into novel groups that change the environment.

Causal maps alone are not enough to build shelters, control fire, and so on. We have had to develop flexible, even novel social species using the same kind of causal sensemaking. Until we started building stuff using novel social structures, all animals changed behaviors in response to environmental pressures. They mutated into new species or learned new behaviors, effectively changing the behaviors of individuals within species—always in reaction to the pressures of the environment. Instead, we have changed the environment to fit our needs because we are able to, by virtue of our application of causal sensemaking to social organizing. We have learned to test the causal maps of social contingencies. In so doing, humans have expanded social organizations from the naturally occurring family and clan groups of other vertebrates to include a host of novel forms invented to moderate environmental demands. We have invented these new social species, organizing ourselves according to "rules" that sometimes successfully, sometimes disastrously, alter the niches we inhabit.

To be clear: Other species have demonstrated the ability to alter their social structures. For example, after pronghorn populations had been diminished in the western part of North America, their social groups were characterized as harems, where the few surviving males had several fertile females to tend. As populations rose, there was more competition among the males, leading to family groups headed by females, with males in more ancillary roles. These changes in social structure were reactive, meaning that the pronghorns learned new behaviors in response to situations, with their social structures following from this (Byers & Kitchen, 1988).

In humans, social structures are sometimes deliberately constructed and altered according to our understanding of what follows from what. We've wound up with a huge variety of social structures as a result, all attempting to alter the niches in which they are put to the test. Social distancing in response to the COVID-19 pandemic is one of many examples of a deliberate, novel social structure. It followed from the old quarantine idea of the Middle Ages in Europe but is not as isolating and doesn't have the draconian enforcement required for this earlier social form. It worked reasonably well because our social maps were "good enough"—a point I return to in Chapter 8 when considering the problem of error.

SUMMARY

A special kind of adaptive strategy has allowed our social species to circumvent the rules of evolution. Instead of reacting to our ecological niche only by adapting individual forms and behaviors, we have altered our niches to try to moderate environmental demands. This evolutionary hack has been made possible by structures (neocortex and social processes) and behaviors (cognitive learning and social learning) that short-circuit the usual rules that pit the carrying capacity in a niche against the forms and functions that usually constrain genetic speciation and behavioral learning. Now that most of the earth is our niche and we appear to be reaching its carrying capacity, we need to figure out how to manage the social speciation hack. The usual four attention–problem identification–deliberate thinking–learning steps help, but because time is short, the fifth option—taking action—is required. We need to aim at least one point faster on the scale of adaptive change.

How do we manage social speciation? To answer this, we need to understand how it happens.

3

Human Exceptionalism

It is easy to agree with Henrich's (2015) argument that human adaptive strategies are different from other animals' strategies. We have done all the same kinds of adapting that other animals have done—genetic change, behavioral learning, imitation, cognitive learning, migration, and even rigid social structures (e.g., bees and ants; Foitzik & Fritsche, 2019) have all been parts of our *reactive* adaptive repertoire. But humans have also taken an *active* approach—using some special cognitive characteristics to alter our social structures, then using these structures to test various alterations to our physical niches. In this chapter, I begin to describe how this active approach makes us fundamentally different from other species. This quickly confronts the moral question of human exceptionalism as we begin to piece together the human structures and functions that have combined into our new, profound, and powerful adaptive strategy. I try to describe the nature and emergence of this human "exception," of the

https://doi.org/10.1037/0000296-003
Sustainable Solutions: The Climate Crisis and the Psychology of Social Action, by R. G. Jones
Copyright © 2022 by the American Psychological Association. All rights reserved.

improvisational social organizing that has yielded entirely new forms of life on this planet.

Human exceptionalism is sometimes used preemptively in controversies about sustainability. The ethical question of whether human interests should take priority over other life is answered with a profound, value-laden "Yes! We must look to human survival first, as a moral priority." But this moral argument misses the point. The human exception is not so much a moral imperative as a matter of demonstrated fact. Our survival has taken precedence over the survival of other species in fact, not as a deliberate prioritizing of humans over other life. Revealing the nature of this "exceptional-ness" from the applied psychology perspective provides important new avenues to navigate a more sustainable future.

When we talk about saving the planet, the question of exceptionalism does take on great moral significance. If we are exceptional, some suggest we have a right to decide what we're trying to save. Do we mean to save just people from extinction (the exceptionalist view)? Of course, this will rely on saving some of the other species on which we all depend (e.g., domesticated animals and food plants). But some people argue for a broader description of what needs saving. So, does saving the planet include saving wasps, mosquitoes, and dandelions, which many people consider "pests"? Again, exceptionalists would say no, let these die, or, if necessary, kill them in order to save our own "special" species.

Answers to these questions even come from ancient religions. Some foundational religious texts state clearly that we are the lords of the world, set up to dominate and control it. If we take this as a moral belief—a view shared by billions of people—then there is a prescription for us to master things in our world. We "must," "should," "ought to," "need to," and "have to" master the world.

Applied psychologists are more inclined to take this instead as a descriptive statement—that we are, in fact, designed to control the world around us, for better or worse (or both). This view suggests that the ancient wisdom makes no requirement, no prescription for us to dominate. Instead, some of these texts provide a succinct description of what makes us different from other animals. We are built to change our world. Our special adaptive strategy is to design and construct our own safe and comfortable spaces using pieces and parts from the natural world. Ancient texts may have understood that we are hackers before we had this word to describe it.

UNDERSTANDING THE WORKINGS OF HACKED ADAPTATION

Even though we have hacked the survival code, we are still human—or rather, still subject to the general rules of evolution. We still adapt like all life on the planet through the survival of certain genetic forms within the constraints of an ecosystem. We also adapt like some other animals by learning from one another, mostly by copying others' behaviors and from trial and error on our own (behavioral learning), as we have already seen. It's just that we're not quite as reliant on these survival strategies as most other species for several reasons.

First, as we saw in Chapter 2, we develop general rules rather than relying on rote trial and error learning. Like a few other species—not coincidentally all of whom have the neocortex—we learn to "map" our circumstances. It's worth noting that none of these species have achieved our population levels or dominance, even though they have this cognitive learning capacity. I haven't seen any bonobo cities or elephant cars, though *The Hitchhiker's Guide to the Galaxy* (Adams, 1979/2007) suggests that porpoises have been in charge all along—we just haven't recognized it. The point is that we are not unique in our causal sensemaking (Gluck & Myers, 2001; Tolman, 1948). Other animals do it, just maybe not quite in so complex a fashion as we do. Still, like us, these species build mental models of the contingency rules of their local circumstances. Like us, these mental models are used to help them get the things they need—and avoid things they need to avoid.

Our Causal Reasoning Machinery Is Imperfect

I use the phrase "help them" because these are heuristic models. Like us, these cortex-toting animals are forced to accept a certain amount of error in their contingency rules. It may help that we have more complex maps than others of our ilk, but it is now generally accepted that our causal maps are deeply flawed. Cognitive psychologists have been demonstrating for decades just how biased our leaps to causal explanation are. There is a long and growing tally of errors, biases, and heuristics that, although they serve us "on average," are wrong in many instances (Cialdini, 2007; Kahneman, 2011). Where careful, deliberate thought could have led us to better choices, we rely instead on biased rules such as hindsight, confirmation, and a couple of dozen other clearly demonstrated shortcuts that tag along with our fancy mental modeling (Jones, 2020).

Confirmation bias is perhaps the best example of our inclination to leap to causal explanations. It shows how we rely on expedient rules of thumb to support our current causal maps. Instead of spending time we don't have trying to figure out precisely why things work the way they do every day, we tend to find a ready explanation for what we see happening around us, then rely on whatever confirming evidence we have at hand (Jones, 2020). This is the opposite of what science and logic do. These disciplines seek to disconfirm other explanations until an explanation that can't be rejected is left.

I learned about confirmation bias from Tom Ostrom, a social psychology professor at Ohio State for many years. Tom would present three psychological findings to his students, asking them to mark down whether they were surprised by each finding or not (in a yes/no note to themselves). He would then ask the students to make note of any explanations they had for each finding. After all three findings had been presented, the students would have a chance to report (a) whether they were surprised by the finding and (b) what explanations they had either supporting the finding or some reason for their surprise.

The findings were drawn from World War II combat infantrymen and went like this:

- Soldiers with more education were more likely to suffer from post-combat trauma than were soldiers with less education.
- Soldiers who grew up in rural settings made better adjustments to military life than soldiers raised in urban settings.

- Before the war was over, fewer veteran soldiers reenlisted than after the war was over.

In many years of doing this exercise in my own classes, I have never had a majority of students report surprise at any of these three findings. Usually, there is a smattering of hands when I ask, "Were you surprised by this finding?" Only a minority said that they were surprised.

Students' explanations for the findings were also interesting. I usually heard a short list of causes: "Better educated soldiers think too much," and "Rural soldiers know how to hunt and kill animals." But there was almost always at least one new explanation—one I'd never heard before. For example, one freshman suggested that soldiers from rural upbringings are excited to learn about the many different people outside their small, rural communities. Creative explanations like this one have always fascinated me, so I will return to them a bit later.

But first, it is important to understand that these findings are the opposite of what was actually found in the large-scale study reported in 1946. Soldiers with more education were less likely to have problems, urban soldiers did better on various measures than did soldiers with rural upbringings, and soldiers were more likely to "stick around" before the war was over than after (Booth et al., 1978; Star, 1949a, 1949b; Stouffer, 1949). Tom and I lied. But we also demonstrated that people don't tend to look for counterexplanations when a professor (or other "expert") tells them a "fact" and that, further, we tend to seek confirming explanations for the findings presented by our authority.

This exercise demonstrates our tendency not to do what scientific disciplines do: disconfirm hypotheses until one is found that cannot be refuted. Understanding this confirmation bias also helps cement the need for a social structure to achieve our adaptive hack. In fact, this exercise was used originally to demonstrate to a group of scientists without credentials in scientific psychology why there is a need for a science of psychology. If everyone can come up with all kinds of explanations for the opposite of reality, there is a need to test these explanations if we want to uncover the actual reasons for our thinking and behavior. But to do this, science needs a lot of hands on deck, not just for generating studies and findings but for peer review of the reported findings, critical replications of studies, and a host of other roles played in the societies of science. This exercise clearly demonstrated the need for a social approach to scientific psychology.

Even more revealing to me was that it demonstrated how quickly students arrived at direct, causal explanations. When asked how long it had taken them to arrive at their explanations, students reported "almost no time at all!" This ability to rapidly connect causes and effects into contingency chains is remarkable, if flawed. Even though we are exceptional, our cognitive reasoning is inclined to errors. We are what social psychologist David Funder described as "dancing bears." Although we dance poorly (have biases in our causal reasoning), we have still managed to develop remarkably effective cognitive maps. To dance at all is a pretty big achievement for a bear (Funder, 1989).

So, let's not confuse our superior ability at causal sensemaking with perfect sensemaking. Far from it. It is not "superior intelligence" alone that makes us so successful (Harari, 2014), especially because other animals have similar abilities (Gluck & Myers, 2001).

Reducing Error in Our Mental Models

Along with disconfirming scientific tests, effective subrules can reduce this kind of error. This often comes with more time spent in the circumstances but also by social learning. Because we live within our particular circumstances, we are able to construct more accurate "A→B, under C circumstances" rules, both by testing our ideas and by watching others do so. Many of us do this kind of map perfecting, at least when we're under the motivational circumstances where such learning occurs (Balzer et al., 1992). The more accurate our mental models, the less error we are forced to accept.

This can be seen by looking at one of the fundamental activities all animals must engage in: getting food. Even the contingency rules for this most basic need—to satisfy our hunger—are not straightforward. In infancy, the rule "If I cry, I get food" works often enough for it to be repeated. Through the eons of contingency map development, humans have learned that "If I shoot an arrow at an animal, I get food"; "If I plant a seed and wait, I get food"; and "If I pay money, I get food" can also be used. But these rules all depend on where we do the crying, shooting, planting, and paying. These very different sets of successful heuristics illustrate how dependent on circumstances these contingency maps can be. Learning which cause leads to what effect under which circumstance must take a lot of time, right? Perhaps so, with the exception of crying because most infants do this without initial learning.

Animals with cortices have advantages here, depending on the same old rule of evolution: The complexity of our mental map matches well enough with the actual complexity of our ecological niche. Form and behavior need to match ecological requirements. But even accurate mapmaking comes with a trade-off: complexity. Following the food example a bit further, anyone who has done the grocery shopping for a while knows that it can take time to develop an accurate mental map of the grocery store.

Like other species, the timing of this approach to adaptation matters, too, and this is the second trade-off. If we learn an accurate contingency rule quickly at the right time, we get what we want. However, if we don't learn these complexities quickly enough, we might use the wrong rule for our circumstances. In the food example, we bring a bow and arrow to a grocery store. Learning which cause leads to what effect under which circumstance often has a time limit. And because it can take a lot of time to learn the right rules, we are naked to the effects of error unless we think fast enough. And we're back to the wonderful expediency of a heuristic: "Bring a wallet to the store" helps—most of the time (Kahneman, 2011).

Most of the cortical animals have solved this time–complexity problem by being sociable. Elephant babies are taught by their successful elders early in life to engage in some behaviors (e.g., mud wallowing to keep pests off themselves) and avoid other behaviors (e.g., running away from the group when others are "frozen" to avoid predators). They can use these heuristic rules enough of the time that they succeed at reaching maturity. Plenty of other species do this, so this is where theories positing superior human intelligence (Harari, 2014) and the passing along of information through generational social learning (Henrich & Muthukrishna, 2021) fall a bit short of understanding what makes us so successful. We are not set apart from other species because of social learning, but it's helped.

The third trade-off from complex mental maps brings us right back to the problems that all species confront. How do we adapt to changing environmental circumstances? If our parents lose a basic food source, crying does us no good. If there is a drought, planting seeds won't lead to much food. And grocery stores have been exiting residential areas in many U.S. cities, so what can people without cars do to get food? These are the types of changes that other cortex-carrying species have been experiencing for a long time. It probably explains the extinction of some of our closest relatives—earlier versions of the *Homo* genus, such as our Neanderthal and Denisovan ancestors. When their ecological niche changed, they were left in the lurch. So, what makes us different?

Old Forms Used in New Ways

Despite these trade-offs, our unique human form of social adaptation does rely on the causal reasoning made possible by the neocortex. Following the flying example of structural change, the neocortex is like membranes that form between limb structures that eventually morph into wings. If we substitute "thin, exposed membranes" and "feathers" (seen in fish and dinosaurs) for "cortex" and "social learning," we can see that animals other than birds have membranes and feathers but are unable to fly. As with bats and birds, the new ways of using membrane and feather structures led to the breakthrough of flying. As with flying, it was the development of other behaviors that used the cortex and social structures in new ways that has led to our current success.

We need the cortex and social learning for our hack. We are most certainly reliant on both in some ways that I have not examined yet. But unique to us and maybe one or two other species is the imposition of our contingency maps on the world around us. When we figured out how to keep comfortable under a roof, coordinated the efforts of other people to help us build a roof, figured out how to heat the resulting space when it was cold and light it in the dark, and so on, we were doing a whole new variety of adapting. Again, we were not changing ourselves to adapt to an ecosystem—we were actively changing the ecosystem itself to make ourselves safer and more comfortable. This kind of behavior is our special adaptive tool.

And it relies, among other things, on testing causal contingency rules on our social structures.

SOCIAL INVENTION

Back to contingencies of getting food. Note that all of these have social components: crying to caregivers, imitating hunting and seed planting and harvesting, and following marketplace norms. All these involve observing and interacting with other people. Crying is inborn, but these other two components started as social inventions—agriculture and market centers were both social inventions that are entirely reliant on coordinating a large number of people's behaviors. Our ancestors had to figure out these new ways to organize themselves to adapt to changing environmental conditions.

Further, we have proven over and again that these kinds of social inventions can be used to change the environment to suit our needs—depending somewhat on which environmental circumstances we're in. We have already proved that we can make our lives better by using free markets and regulations on free markets, inventing new technologies and reverting to old ones, and legislating government-enforced regulations and removing such regulations. All of these, large and small, are socially derived, dependent on the ability to flexibly reorganize our social groups. We wouldn't have been able to do any of them without learning from others, working cooperatively with others, buying and selling from others, and so on. But we also wouldn't have done them without trying out new social structures in the first place.

Proof of the success of this socially derived adaptation lies in the fact that the daily lives of most humans today in most places on earth are better than the lives of the wealthiest and most elite people of any time before the 19th century (BBC, 2010). Again, we live longer, healthier lives; have many more and more sophisticated amusements; work under safer, more humane circumstances; and generally live the most comfortable individual lives of any humans ever. There are also many more of us (and many more of us to come), which is one measure of our adaptive success. This is all because of the socially derived adaptations that we have invented by enacting our mental maps—by social niche building to hack evolution.

In his brilliant analysis of human adaptation, Harvard biologist Joseph Henrich (2015) zeroed in on something he called *collective intelligence*. He made reference to cooperation (which requires language) and experimentation with social norms as the processes through which we have molded the world to fit our survival needs. His main underlying mechanism is *culture*, which is knowledge passed on through groups over generations. He argued that our adaptive advantage is derived from cultural learning, that we have altered the world so thoroughly because of the knowledge we share across generations about how to organize and behave in the face of our local environment's demands.

I have already described how we develop heuristics, which can be learned through social observation. We also saw that cognitive learning occurs more readily when we are in groups and that we can learn by watching and testing the cause-and-effect relationships we think we have observed. This is what would usually be called *social learning*—a process of cognitive change that occurs more readily in groups (Ford, 2021; Woodward, 1982). But although this describes one of the mechanisms through which we have formed the social groups needed to make environmental alterations, it does not explain the specifics of this process (cf. Burgos-Robles et al., 2019) and especially the motivations underlying it. Risk aversion, opportunity seeking, and another kind of cognitive change all need to be accounted for if we hope to manage our inclination to alter the world around us.

I suggest a different term than Henrich's *cultural learning*. I prefer to use the more precise term *shared mental models* (Coovert & McNelis, 1992; Klimoski & Mohammed, 1994; Mohammed & Dumville, 2001). There are several reasons for choosing this term. First, it is easier to demonstrate than cultural learning. One way to demonstrate shared mental models is by thinking about common instances when our cause–effect expectations are contradicted. Part of the fascination of a magic trick and the joy of a joke is that both of these break our cause–effect expectations. Consider the magician who makes a quarter disappear into thin air, then recovers it from behind a child's ear. Even young children are a bit awestruck by the inconsistency of these events with their usual expectations. These expectations have more to do with physical reality than with culture, however.

An example that does encompass an aspect of culture (in this case, language) is puns, which use a word with a common (cultural) meaning in a way that is not consistent with this common meaning. For example, the word *boot* can mean something other than sturdy footwear. Consider this statement of fact: "John gave Jim the boot." For most English speakers, the phrase suggests that John was Jim's boss and had fired Jim from a job. But if we include the picture in Figure 3.1, our expectation is that the word was used for another meaning.

Puns can be found in many languages, not just English. Hence, contradicting expectations does not rely on culture, even when shared expectations are described using language.

Understanding this contradiction of expectations also illustrates the conditions for learning them. Magic and humor only garner surprise and laughter under these conditions. First, as Henrich suggested, they occur

Figure 3.1

"Give him the boot!" Illustration by Marty Two Bulls Sr., Santa Fe, New Mexico, United States. Copyright 2021 Robert G. Jones. Used with permission.

almost by definition when someone else is doing them. We are surprised and laugh, not when we do the magic trick or tell the pun but when someone else presents them to us. Because these show that the "expected" cause–effect relationships that we've learned do not hold, we learn that there are exceptions to the A→B rule, at least in the presence of certain other people with whom we share expectations (for cultural or other reasons). By extension, we learn and change in the presence of others—we are fundamentally social beings. On this, Henrich agreed with most psychologists (Conway et al., 2009; Feng et al., 2021; Woodward, 1982).

Second, magic is surprising and humor gets a laugh when we share some set of social expectations. My quibble with Henrich's use of the term *culture*, which he posited as the essential basis for our success, is that it is a broad term applied often to large, long-lasting groups of people. Not surprisingly, then, anthropologists have included many things other than broadly shared expectations in their definitions of culture. Anthropologists and psychologists also define culture by referring to artifacts, everyday activities, rituals, and values—not just shared social expectations (cf. Harris, 2019).

Psychologists have defined shared social expectations using the more precise term *shared mental model* (Coovert & McNelis, 1992; Klimoski & Mohammed, 1994; Mohammed & Dumville, 2001). I prefer this term because it accounts for shared thinking that is not necessarily part of a broader culture. Applied research starting in the 1970s identified shared mental models (and related concepts—i.e., cognitive models, mentalizing networks, neural networks) as important bases for behavioral expectations in corporations, music ensembles, and other social groups (Arioli & Canessa, 2019; Bougon et al., 1977; Coovert & McNelis, 1992).

Third, and most important, stable culture is not what gives us the power to adapt. We do not socially speciate because of culture but despite it. I argue that the kind of cognitive change that occurs in the presence of other social groups is developmental change—not the learning type of change passed on through culture at all. We change the fundamental expectations that we have for ourselves and others. Cultural expectations help to maintain these shared mental models through social learning. But our hack relies on altering these models.

WHAT MAKES US SPECIAL: SOCIAL SPECIATION

One of the primary conclusions from an analysis of human adaptation is that we are fundamentally social creatures—that our species is perhaps best thought of as a collective rather than a population of individual creatures. Most of what we have done to adapt differently from other animals relies on our ability to use cognitive mapping to improvise, mimic, construct, and change existing social organizations. Thus, if we are to manage the aspects of our species that have made us both so successful and so destructive to our environment, we need to focus our efforts not so much on individual learning and change as on this social construction process. Individual attention, problem identification, deliberation, and change are all susceptible to the social speciation process.

Back to exceptionalism: We do, in fact, dominate our environments, whether we should do this or not. A core conclusion I have reached from studying and applying psychology is that we are "exceptional" by virtue of the way we deliberately alter the social world to adapt it to our needs. The next few chapters examine the cognitive and social processes through which *Homo sapiens* have created what amounts to a new, social life form.

4

Psychological Factors Required for Social Speciation

Archaeological evidence strongly suggests that humans have lived in small family or tribal groups for a long time (Jones, 2020). These primary social groups are certainly alive and well today and, in some instances, are still highly adaptable to changing local conditions. In fact, the continued success of families and tribes probably depends on their flexibility—their capacity to deliberately respond to demands posed by the many environments in which they operate (Masuda & Visio, 2012). This includes adaptation to both physical and social environments, but we will be concerned mostly with social adaptation from here on.

Industrial and organizational (I/O) psychology and its related fields have already developed and implemented some applications to manage such social adaptation. These efforts have targeted and relied on "primitive" social behaviors that can be seen in other species. On the "I side," we try to manage the tribal inclinations that lead to gender and ethnic discrimination, nepotism, and cronyism. On the "O side," we have developed

https://doi.org/10.1037/0000296-004
Sustainable Solutions: The Climate Crisis and the Psychology of Social Action, by R. G. Jones
Copyright © 2022 by the American Psychological Association. All rights reserved.

complex feedback interventions that rely on the social nature of learning and motivation. These I and O approaches are relatively easy to justify to client organizations and have found fertile ground due partly to their effectiveness.

However, to manage deep adaptation and social speciation for the benefit of sustainability, we need a broader understanding of the distinctly human processes that lead to shared mental models, including social identity, creativity, and developmental change. There are some scientist–practitioners who have taken up these challenging constructs for other purposes (e.g., Dono et al., 2010; Hernández et al., 2010; Markus & Kitayama, 2010; Patrick & Hagtvedt, 2012), but it is fair to say that they have not been in the mainstream of psychology in organizations. Argyris's (1982) idea of double loop learning approaches this meta-adaptive idea, as well, only without the specific variables used here (identity and development). But serving as meta-adaptive feedback providers during social speciation is close enough to current scientist–practitioner competencies already that it provides a promising approach to successful, sustainable future adaptation. I think we will find that a more thorough understanding of the contexts, capacities, and especially motivations that lead to social identification, creativity, and developmental change will serve I/O psychology well, regardless. In the meantime, they are proposed here as key variables in social speciation.

SOCIAL IDENTITIES

In Chapter 1, I argued that change in social identity, where people see each other as belonging to the same group, is sufficient but not necessary for realizing social speciation. Still, social identities do play important roles in the ordinary course of a workday (Ashforth & Humphrey, 1993) and the important process of steering toward new mental models (cf. Patrick & Hagtvedt, 2012; Sanchez-Burks et al., 2015). Even though familial and tribal impulses still exist, we have developed many other social structures to help us survive. These often take on a life of their own, establishing new social identities (e.g., fans of sports teams, brand loyalists, commitment to organizations) and retaining old social orders and social roles. These tend to stifle social innovation (Sanchez-Burks et al., 2015) and protect people from the kinds of conflicts that spur developmental change, so they can be important barriers to social speciation. There are also social forms that have pretty much fled the scene or morphed into other social structures. An example is the monastic system, whose function of preserving knowledge has been handled more efficiently by other, more recent social species (e.g., libraries, universities, Google).

Applied psychologists have spent most of their time managing forms such as family and tribe that have survived over long periods or forms that have become dominant in the relatively recent past (especially work organizations, justice systems, and commercial enterprises). There is an almost endless list of these latter forms, which is also testimony to how different we are from other species, whose social structures tend to be quite straightforward, take on a limited number of forms, and are often based on rigid social roles. Our social variability is remarkable, so understanding what "works" is a likely avenue for successful future adaptation. What makes some social forms fade from the scene while others thrive?

Survival of the (Social) Species

One answer comes from understanding the purpose of a social structure. We have devised these social arrangements to solve various problems. Some are the "survivor" forms commonly referred to as traditional social forms—in addition to family and tribe, we have armed forces, markets, governments, religious groups, guilds, and so on. Others are local variants on broader forms; pubs, bars, and clubs are all local variants on drinking halls that started as venues in marketplaces to exchange information. What they have in common is that almost all help to keep nature at bay one way or another, mainly by helping people learn how to organize in the face of various adaptive pressures. The farmers at the market pub and the commodity traders at the downtown club are all trying to find ways to organize their activities to the best advantage in the face of challenges. Most of these social forms are socioevolutionary hacks that have worked well enough for long enough that they are retained as potentially helpful. Few are actually novel forms, at least after thousands of years of existence in various shapes and sizes.

Then there are the exceptional social forms devised deliberately to solve a novel set of problems. Markets and pubs were examples of these novel forms, at least when they first appeared on the scene. Rather than trying to barter with the close neighbors, farmers across a small region banded together to create a central clearing space that yielded a broader range of items for exchange—and market towns were born. Places sprang up in these market towns where people could meet, relax, barter, and learn from one another—whence pubs.

As socioevolutionary hackers, we have come up with new, deliberately designed social structures to reduce our susceptibility to the whims of genealogy, local climate, large distances, and other adaptive problems. Important examples from the past 1,000 years or so include professional societies, universities, hospitals, bureaucracies, insurance cooperatives, and international cooperative organizations. We do not yet know much about the psychological processes through which these exceptional structures occur, but there is a fairly recent example from which some hypotheses can be derived.

Engineering as a Novel Social Species

Arguably, the most influential of these social species over the past 2 centuries is the society of professional engineers. In his brilliant book *Rising Tide*, John Barry (1997) described the founding years of this species—a highly enterprising, somewhat organized group that applied physics to controlling forces of nature that had dominated human affairs forever before them. Barry's description of one of the most famous of these early engineers, James Eads, demonstrates the importance of social invention to engineering. Above all other aspects of his character or situation, Eads's "ability to avail himself of the skill, of the experience and the brains of all with whom he came in contact, was phenomenal and enabled him to succeed in any mechanical proposition suggested" (p. 51). Eads was one of the founders of the engineering profession—a social species whose impact on daily life has been so profound that parts of both the physical and social world would be unrecognizable to someone who lived before it emerged.

What is the odd combination of factors in our nature that has developed into this capacity to successfully alter our social environment? I suspect that fundamental family and tribal social groups are successful partly because they allay people's fears. Risk aversion weighs heavily into the continuation of such groups (Alessandri et al., 2018; Guassi Moreira et al., 2021), which makes it likely that they will not go away. But what of other, more deliberate social construction projects? How and why have we organized in so many ways?

Just as with the flying example of adaptive change, the structures of change and the behaviors that go with these structures have coincided in our species. Stepping further into our neocortical thinking, I argue that we are endowed with a unique combination of cognitive drivers that take us well beyond just causal reasoning. We (a) have a tendency toward active curiosity and (b) are endowed with structures in our head and neck that allow most of us to use our brains to control communication. Although these factors come in varying degrees across our species, they are extant enough to make our special kind of sociobehavioral adaptation possible.

CURIOSITY AND REALITY TESTING

We have already visited causal reasoning's origins in the gradual evolution of the neocortex, a structure that has been around for quite a while—long before our species figured out how to apply it on such a large scale. Alongside the causal mapping ability that came with the neocortex, humans have also evolved the means and inclination to test our mental maps, including the mental maps of our social relationships.

There are other species that demonstrate both curiosity and an outcome of curiosity—playfulness. As with other great apes, our curiosity inclines humans to test the contingency maps we have constructed in our minds. When things seem safe, we go to the trouble of testing these maps in reality. We look to confirm that the causes we ascribe to things are, in fact, the causes at work in the world. In children, we call this play. In grownups, it may be play, too, but "inventing," "creating," and "experimenting" are all descriptive in different contexts. In his popular book, *Sapiens*, Noah Harari (2014) described these briefly as imagination and storytelling. A few other species get playful, but we've made it into our daily activity. We have even applied this curiosity to devise an enormously successful system for testing whether our causal explanations are real, calling it "science." Engineering is a dramatic attempt to apply science in a socially organized fashion to manage risk.

The science of psychology has even identified the characteristics of individuals who are more or less inclined to test causal explanations. One of the most influential personality characteristics identified by psychologists in the 1980s as part of the Big Five (Costa & McCrae, 1988, 1989) is one that most people have not heard of, but which is only slightly less prominent than the better known dimensions of neuroticism and extroversion. Openness to experience is an important individual difference variable. It is defined partly as a greater or lesser inclination to explore, try new things, and be intellectually curious (Zare & Flinchbaugh, 2019). Unlike other personality characteristics, it is the only one that correlates with mental abilities measures (Judge et al., 1999). The theory here is that children who are encouraged to be curious will get better at learning than children who are either ignored or punished for asking the kinds of questions most children ask (Jones, 2015). When a 3-year-old asks, "Why is the sky blue?" their caregiver may reply, "Shut up; I'm busy" or "Not now, dear." But the child who is reinforced with some kind of meaningful explanation ("It's blue because we see light refracting through certain molecules in the air") learns that (a) asking questions can yield positive attention, and (b) there are meaningful explanations for things they observe. They also learn that it is sometimes safe to test causal explanations because such reinforcement increases the likelihood of repeating their initial questioning behavior.

Curiosity Testing Is Reliant on Our Social Environment

Because of this kind of reward history, many of us are driven to test our causal maps. We take gambles to see whether we can "win" a game, start business ventures to try to make money, try on different clothes to see whether we can attract the right kind of attention, run scientific experiments, and so on. We are cognitive explorers, testing the accuracy of our mental maps.

Our ability to try out these ideas in the real world relies on a huge variety of prior social inventions. For gambling, we need someone to invent a game (usually not us) and often need someone to play the game with us. For businesses, we need customers and a social marketplace of some kind, not to mention the equipment manufactured to produce and transport what we're selling, builders for buildings and websites, language we learned from others to communicate what we're selling, and so on. I might figure out that having something over my head keeps my hair dry in the rain or, for that matter, that the neocortex is essential to causal reasoning, but my ability to test this knowledge is limited by the social structures at my disposal. I need help building shelters, and I need a well-equipped lab to try out my brain theory.

Our curiosity testing sometimes looks pretty solitary: a child playing with bubbles, a scientist at her bench, a composer working at a keyboard. If we can do all this good creative work without others, why do we need other people? How is our hack necessarily social?

One answer is brought to us by biologists and anthropologists who point out how poorly equipped individual humans are to get food, protect themselves from predators, and so on. We need each other, they say, for basic kinds of survival (Wilson, 1998). This is consistent with the psychologists we've already visited, who have shown that social groups are our "safe place" for learning (Ford, 2021; Hughes et al., 2020). And what are discovery and creation but aspects of learning?

Another, not mutually exclusive, explanation is that we are motivated to do all kinds of things for each other. Many inventions are the result of the empathic frustrations, hopes, and fears we feel for a loved one. There are plenty of examples of inventions designed to help ease a child's pain or increase the leisure time available to a beloved spouse. There are certainly examples of scientists finding cures for their own maladies, but I suspect we'd find most of these cures are motivated at least partly by the need to serve others.

There are even some researchers who believe that we are genetically predisposed to help others, but not just any others. Colarelli and Arvey (2015) found evidence that even in work organizations, our prosocial behavior is most often directed to helping our kin. This does not change the more general fact that we do creative and inventive things for each other—not just for ourselves.

Creativity Is Social

Back to bubbles, experiments, and pianos: I challenge you to think of any creative activity, curiosity driven or otherwise, that doesn't require other people for it to happen at all. For a child to play with bubbles, someone else (usually an adult caregiver) provides the child with detergent-laden water in a small tub, then shows the child how to use a plastic ring to blow the bubbles. Hours of (sometimes solitary) enjoyment ensue but only because people other than the child manufactured the detergent, tub, and plastic ring then passed along the knowledge of how to brew soapy water, dip the plastic ring in it, and blow with just the right amount of pressure to make a bubble. This is to say nothing of a scientist's bench or a musician's keyboard and the extensive learning from others that was necessary for these activities.

SOCIAL SPECIATION

This same kind of combining of different human activities has turned sociability into a necessity for survival and adaptation. We need each other to put our ideas to the ultimate test—altering the world around us. Just as "useless" webs between early rodents' fingers turned into membranes and wings in bats, so our curiosity and social nature have turned into something different: novel social groups that operate on the world, that make our evolutionary hack—our flying—possible. We have developed new social groupings to put these ideas to the test, what we have dubbed *social species*.

Although I will continue to try to define the term *social species* succinctly, it is important to understand that it is the process through which we form novel social groups that matters—social speciation matters more to our hack and to managing this hack than does the idea of a social species in itself. That said, a social species is defined as a type of social group that has been formed to try to alter the environment to moderate its effects on individual human comfort and safety. Social species rely on some members of the group sharing a mental model of the contingencies of the environment that they are attempting to moderate. *Environment* here refers to physical and social forces, the extremes of which are thought to have a damaging effect on individual comfort and security.

Historians, anthropologists, and sociologists have all described social species, presumably without calling them this. I/O psychologists and their colleagues in organizational behavior (a closely affiliated field found in business schools) have developed some ways of testing related theories from these other fields and have identified and invented new social species as a result. A couple of the more well known of these are business organizations that profit by novel storage, organizing, and communicating of social information.

Google's founders set out to create something like a super-encyclopedia of topical information (A. Fisher, 2018) based on the newly organized Worldwide Web (arguably a new social species itself). It was not a coincidence that they used pieces and parts of other organizing methods to cobble together a novel social arrangement for managing their enterprise. They constructed an "open office" concept that resembled more an art studio than a corporate bureaucracy. They encouraged "playfulness" and explicitly allowed for error, both of which would tend to reduce the tendency toward risk-averse, heuristic activities. By enhancing creativity in these ways, they have created an enormous, highly successful, geographically dispersed social species that is being imitated in other workplaces—sometimes to good effect.

Facebook has also been developing a social species through its platform, for better and worse. Although its corporate model may not be especially novel, the platform itself provides a new way of connecting with family and friends through informal, often physically remote groupings (Phillips, 2007). It also allows the user to quite efficiently discover the whereabouts of people with whom they had earlier social connections and integrate them into the novel social grouping created by the platform. And yet another social species has been created just in the past couple of decades. Never mind the millennium or two required for a complex life form to develop into a physically distinct species to match the demands of some small ecological niche—we can alter our social organization on a massive scale in a matter of months.

Safe Places

It is important to recognize that social groups such as family and tribe often provide low-risk environments already—without any new social speciation. Knowing that other members of the group are out on watch, have successfully gotten food, and otherwise allowed individuals in the group to be calm leaves some members of the group able to contemplate, learn, and formulate mental maps at their leisure (described in Chapter 2). More than just being safe, though, family members may be more likely than others to give new ideas a whirl in the interest of keeping social relations happy. In fact, most new businesses start this way—as family affairs (Nicholson, 2015).

There is a more general principle at work here. Anxiety can be death to creativity (Byron et al., 2010). Historically, societies under stress are not places where creativity thrives. Creative thinking thrives in munificent places. History is full of examples such as the Medici period in Italy, the early Song Dynasty in China, the Elizabethan period in England, and post–World War II in the United States, where new discoveries and achievements blossomed. Conversely, Revolutionary China, the early Soviet Union, and the frontier of the Western United States were not known for their creative output. Safe social structures were hard to find in these periods.

I think most people intuitively understand that the usual reaction to stress is to try to reduce it and that this leaves less time for play and other creative activities. Also, a supportive social group has been shown over and again to be an important buffer against the effects of stress (Gonzalez-Mulé et al., 2021; Kim et al., 2018; Mathieu et al., 2019). Our creative social organizing itself relies on the presence of a safe social group.

This posits a self-perpetuating component to this success. Curiosity, cognitive mapmaking, and sociocognitive exploration have created relatively safe artificial niches. These socially constructed niches are often places of actual safety. These social groups are not always safe, either, as when someone comes along with a gun powder cannon to blow down the stone walls of your castle. But at least for a time, the castles and the other safe environments we have constructed have stimulated exploration.

Learning to control this creativity better in the interest of future humans requires understanding the tools and processes of social exploration. What sorts of social tools have been required to translate cognitive–social forms into regular behavioral use? Following the flight analogy, what is the air on which our social speciation flies?

We Need Language

One answer to this may seem obvious. Language has been essential to our adaptation. If we go back to the beginnings of our species, language was probably used to coordinate hunting and defense, teach young group members life skills, and share or trade wisdom with one another. But how it has helped with our enormous hack is not so obvious.

Psychology and anthropology have shown that there are other species with what appear to be advanced communication abilities. Some of these languages happen outside our sensory capacity to detect them. Dolphins and porpoises communicate in supersonic whistles, chirps, and squeals, and larger cetaceans (whales) and elephants also use subsonic sounds to warn, show sexual interest, and teach their young (Marino, 2002). These are all social animals. Our similarly complex use of language shows again that we accomplish things not as individuals but as groups.

To answer the question at the beginning of this section, we need languages for coordinating our complex hacks. Unlike any other animal, we have figured out how to communicate with more than just sounds and gestures. We have learned to write, read, and even communicate in abstract ideas using processes such as computer code. Sciences use highly complex, abstract languages that have no sounds or gestures—mathematical systems specialized to serve their special part of the human hack. Languages have become an increasingly important tool for adapting the social world to meet our needs.

We have also elongated the period that our recent ancestors dubbed "childhood." By keeping younger people dependent on their elders for longer than other species, language can be used to pass along to offspring increasing amounts of information about the world. This is where Henrich's (2015) idea that culture is the "secret of our success" holds some water. But it is the imaginative processes that arise in a successful, safe culture that are essential to our hack. We need effective prior hacks, passed along as culture, to put air under our wings.

SOCIAL STRUCTURES AND THE BEHAVIORS THEY EVOKE

The behaviors we have used to "fly" with these three abilities (cognitive learning, curious testing, and advanced communication) are used mostly in the service of social coordination. We have already seen how a fairly simple form of social coordination has led to building roofs. We will probably never know whether roof building relied on a novel social structure because it happened before recorded history. But there is enough historical record to get a pretty good idea of the novelty of the essential social structures that have led to our current situation.

Scientific Social Speciation

The testing (science) and manipulation (engineering) of causal structures rely on the dedication of resources to special social organizations. In fact, both science and engineering rely at their foundation on invented social structures, including everything from ancient academies and libraries to more modern universities, royal societies, and think tanks. These novel social organizations popped up in ancient Greece, China, Mexico, Iraq, India, and Northern Europe, leaving records of some of the most complex and novel social structures ever devised. These solved what seemed at the time to be impossible problems.

Modern science still relies on a large number of complex, intertwined, and deliberately designed social structures. Scientific theorists, usually trained in advanced languages at publicly funded institutions for higher education, spend time identifying new questions based on previous evidence found in specialized sources (journals, professional societies, informal interactions with others in their field). These theorists state formal hypotheses based on their notions, then either test these hypotheses or hand them off to other trained people to test. The organizations where they do this (commonly universities and government labs) usually encourage the research scientists to report the results of these tests, handing them to peer reviewers from other organizations. Depending on the results of these peer reviews, journals and professional societies provide opportunities to disseminate the findings so that other researchers can then identify the next questions to be tested. Science would simply not have been the force that it is and, in most instances, would not have been possible at all before the invention of these behavioral–social structures.

It is especially relevant that one of the problems addressed by the social process in modern science is confirmation bias. Much of science nowadays has integrated psychological knowledge of our heuristic inclination to confirm rather than disconfirm our hypotheses (Cole, 2013). Rather than giving our cognitive maps free rein to build, the social species of science has integrated this self-correcting element into the system. Because individual scientists' brains are set up to confirm their own pet ideas, the social structures of science are set up to disconfirm these same ideas. To the extent that these social structures have succeeded in this, they provide an important clue. One avenue toward sustainability could well be figuring out how to integrate a self-managing check on the inclination to build social structures that support individual comfort at the expense of other human interests.

The Clash of Social Structures

The advantages this cycle of discovery has brought are so many and sometimes so fundamental that it is impossible to catalog them all. Only (socially designed) libraries and databases can do this. There are, however, some ways to illustrate how essential these structures have been. Their value to our quest for ways to manage social speciation will become clear from some examples.

One of my favorites is David Hothersall's (personal communication, January 24, 1989; 1995) description of what happened to scientists before these structures were well established. There is almost no better illustration of the importance of science than the violent resistance that this particular social invention has received, mostly at the hands of social structures founded on things other than the reality checks that scientists pursue. These include famous examples of the demise of early empirical thinkers at the behest of existing social structures that were threatened by what these thinkers had found. Perhaps the most famous is the death of Socrates at the hands of the elders of Athenian society. Then again, the Western Christian church forced Galileo to publicly recant his finding that the Earth revolves around the sun. There are many similarly dramatic examples. Among other things, these incidents point to how radical and potentially powerful scientific social structures have been.

Another way to illustrate the importance of novel, invented social structures is to consider why the Allies won World War II. There has been considerable debate about the deciding factors of this outcome, and because it is in the realm of history rather than science, there can be no completely definitive explanation. But there are a couple of things that certainly made a difference. The first was breaking the German secret code by the Allies before the United States entered the war. This was accomplished by throwing a group of talented mathematicians, code breakers, and scientists together with the expressed purpose of breaking the code. The second was the completion and deployment of atomic weapons by the Allies before the Axis powers could do the same. The histories of both of these accomplishments are full of examples of diverse social groups that were deliberately designed to elicit creative solutions to well-defined problems. Historians may have other examples of early think tanks—the point is that these succeeded, while groups used for these problems in the authoritarian Axis countries were unable to yield similar results, despite the availability of similar intellectual capital. They were relying on older, less deliberatively defined social species to do the work on a new type of task.

Going back to Google and Facebook, we can see how these for-profit organizations have been deliberately devised to invent new social species. Their success has relied heavily on this experimenting with social structures to solve novel tasks. Some even rely on devil's advocacy, assigning a person to try to disconfirm the ideas put forward for changing these organizations. Because these are new social inventions, it remains to be seen how and whether other social structures will change, integrate, conflict with, and otherwise react with them over time. It also remains to be seen how, whether, and in what form they may survive.

When Circumstances Allow

There has certainly been luck involved in these examples of deliberate social speciation, too. And there are some other circumstances that may play a role in the success of social speciation.

For a start, humans have gone through periods of dramatic social change, where one dominant type of social structure has been all but completely subsumed by another. For example, at the end of the last glacial age, many societies made a transition from nomadic to agrarian subsistence, eventually leaving behind most tribal groupings in favor of city-states and nations. These novel forms are saved for times when we simultaneously (a) have big problems to solve (e.g., invading Mongols, diseases) and (b) have some slack resources to avoid big risks for the moment (Morris, 2010).

Not coincidentally, these are times when individual developmental change occurs. Individual developmental change often happens as a result of conflicts in relationships with others. I examine this relationship between developmental change and social speciation in the next chapter.

DESCRIBING THE PROCESS OF SOCIAL SPECIATION

How exactly does our human adaptive strategy unfold? Now that it is clear that calm testing of cognitive maps by the use of complex social group communication is required, we need to understand the broader circumstances under which these occur.

Jared Diamond (1997), in his Pulitzer Prize–winning book *Guns, Germs, and Steel*, provided an important hint to answer this. Ironically, it seems that our tendency to run into each other's tribes and end up in conflict is also where our most valuable adaptation occurs—it is two sides of the same coin. When we run up against other tribes, Diamond said, we must adapt to their ways of warfare or die. This, he argued, is the motivation for us to develop technologies that will defeat these other "enemy" tribes who may otherwise (and may anyway) pass along the "germs" that are part of his book's title.

I argue that along with "guns and technology," these interfaces between societies lead to interactions that create our new kind of speciation. We adapt by organizing novel social groups. Following Diamond's (1997) line further, we do this to build the technologies (steel swords) and physical structures (fortresses) that are required to fight and protect ourselves from enemies. But we also do it to help fight germs, create cooperative markets, and solve a host of other nonwarlike problems. We use social speciation to reduce the extremes of the pressures we would face without these same structures—including armed conflict, for example.

However, social species do not start with physical inventions (e.g., fortresses, steel, swords). As we have proceeded along our adaptive path, humans have certainly built physical structures—buildings to protect scientific experiments, assembly lines to fabricate medicines, and so on. Again, these are part of our hack. But all these physical changes started as people's thinking about and sharing ideas about what will work. We shared mental models among ourselves and somehow organized into groups capable of fabricating the materials needed for such things as particle acceleration, moon flight, and other advanced activities. We have seen that all these physical changes required certain kinds of thinking and specialized social structures first. We invented and tested social structures to create a long list of social species beyond family and tribe.

In the next two chapters, I take a deeper look at a process through which this occurs and that can be harnessed to manage the construction of social species to control our hack.

There's an interesting paradox here: Most people don't understand many of the mental models that support our hack. For example, few people have expertise in organic chemistry, though this field is essential to modern medicine. Because the social species needed to accomplish medical feats are not obvious, there is a need for manifestations to remind people of their nature and importance. So, we use symbols, from the caduceus staff of medicine to architecturally imposing medical school buildings, to represent this important social species. These kinds of constructions are designed to remind people in a particular society of the sociomental model they share. I suspect many readers are thinking of church spires, capitol rotundas, business skyscrapers, university quadrangles, and the like, and you're right—these are such structures, built to remind people of some shared notions about what is "sacred," who holds power, and how we succeed together.

SUMMARY

Our core adaptive strategy, then, is to create social species to support the construction of evolutionary hacks. These hacks allow us to change the world to meet our needs, creating a cycle in which deliberate alterations in the environment increase individual comfort and safety, which leads to creative thinking and the testing of creative ideas through social speciation and further altering the environment. Again, rather than adapting by changing the structures and behaviors of individual humans, we have gone to great lengths to avoid doing just this. Some of us have figured out how to build fortresses, make steel, and so on, as ways to not have to adapt in the old ways that other animals rely on entirely. By changing the physical world to meet our needs, we have avoided selection pressures and expanded far beyond the usual carrying capacity of most other species. We have built novel social species to support these efforts to avoid natural selection. Even more radical, because we are so dependent on these social structures, we have built reminders of our social species in the form of institutional structures.

If this were not enough, we are now confronted with the need to build new social species to reduce our impact (hopefully without simultaneously leaving ourselves more vulnerable to selection pressures). Further analysis of the core adaptive strategy of social speciation will show a path to successful future adaptation. Using meta-feedback to help guide the speciation process, formulating the shared mental models that serve as the outcomes of these processes, and helping to design the system to address its main levers (social identity, creativity, and developmental change) are the main roles that scientist–practitioners can fill in the interest of a more sustainable future. But first, we will see how the creation of new social structures is driven by motives for individual and small group success at the expense of broader, species-level adaptation. The path to successful future adaptation follows from an understanding of (a) the scientific findings about the ecological factors that lead to speciation, (b) applying this knowledge to our social speciation strategy, and (c) developing an understanding of the social and psychological factors that constrain this path to social speciation.

5

Social Margins, Conflict, and Developmental Change

The individual characteristics that support social speciation—curiosity, imaginative mental mapping, dependence on sociability—are like the feathers, membranes, light bones, and streamlined physical forms that support some species' attempts to fly. But to understand and, ultimately, to better manage social speciation processes, we need to understand the conditions under which risky social organizing "takes wing." Motived responses to situational forces certainly lead to change in individual behaviors, as in behavioral learning, which has been a focus of much environmental psychology research over the years (Jones 2015). But we are concerned here with the risky, complex changes in social organizing that, while individual responses to the situation may provide leverage, go well beyond incremental behavior change (see Snell-Rood & Steck, 2019). Changing the mental model shared by a group is a more challenging motivational problem. In this chapter, I address this challenge, mainly by

https://doi.org/10.1037/0000296-005
Sustainable Solutions: The Climate Crisis and the Psychology of Social Action, by R. G. Jones
Copyright © 2022 by the American Psychological Association. All rights reserved.

examining the situational forces that elicit developmental change in one or more members of the group. The forces that stimulate change in an individual's mental model are also important for managing the processes through which shared mental models are formed. Harnessing social speciation may rely partly on knowing how to stimulate, dampen, and otherwise direct this more complex kind of change.

Scientist–practitioners have little control over the characteristics of the individual clients with whom we work. Even if our clients include sociable, curious, imaginative people, broader organizational circumstances can make an enormous difference in the success of interventions (J. W. Hedge & Pulakos, 2002; Mullins & Cummings, 1999). As a result, finding ways to manage social situations to elicit new behavioral patterns (and mental models) is often a central challenge. In fact, high-impact interventions often rely on efforts altering social circumstances through various means, such as management development, family therapy, reframing policing models, and altering work group processes. To manage this kind of change, scientist–practitioners need to understand developmental change and, in some cases, more directly how shared mental models develop and change. Understanding developmental change as a process rather than a state of being (Jones, 2015) allows for its effective management.

Current research provides some evidence about the situational factors that trigger and sustain this core aspect of social speciation processes. For a start, as with the transfer of learning, people are less likely to test new approaches when things look too risky (Blume et al., 2010; Ghosh, 2014; S. Liu et al., 2014). When we are safe—when we are confident enough of our ability to cope with the consequences of risky new behaviors—we try leaping into our sort of flight (Ghosh, 2014; S. Liu et al., 2014). Although heightened risk perceptions can also influence behavior (Sheeran et al., 2014), relatively safe conditions give wing to our curiosity, imagination, and social improvisation (Byron et al., 2010; da Costa et al., 2015).

What are the conditions that make us feel safe? Again, we have a relevant answer to this: Social support gives a big boost to learning (Ford, 2021; Ghosh, 2014; Hughes et al., 2020; S. Liu et al., 2014). Not surprisingly, we tend to test our ideas in the real world with groups we trust

(Klasmeier & Rowold, 2020; Klimoski & Karol, 1976; Li & Hsu, 2018), though this relationship is complex (W. Tsai et al., 2012).

But people are motivated to change the world around us so that we don't have to change our own individual behaviors. Unlike individual learning, if we can organize ourselves to alleviate an environmental stressor, we don't have to change our individual behavior in response to this environmental stressor. By successfully altering the environment, we are freed from such day-to-day pressures to change. The combination of curious cognitive map testing with languages enables us to coordinate group behaviors for this purpose.

This, paradoxically, allows people to behave more autonomously as individuals. So, we are motivated toward social speciation because it makes it possible to moderate environmental pressures—to keep them in a range that reduces their more extreme effects. And by moderating environmental pressures, we lower the individual stress that comes with uncertainty about these extremes. More than this, lower stress allows for the creation and testing of further alterations. If we can hold local pressures within some reasonable range, we have greater freedom to explore, invent, and test our next inventions (Bühler & Nikitin, 2020).

This cycle of reduced pressure→creative thought→social building→ reduced pressure, is illustrated by the following example.

STAYING WARM—AND COOL: THE STORY OF EVOLUTIONARY HACKING

Adapting to the temperature of our local environment has led to all kinds of both basic and learned responses long before humans came up with heated shelters, air conditioning, and other complex social inventions. As with all other species, genetic changes, such as developing more or less skin surface, led some people to hold heat inside their bodies more effectively, while others circulated heat out of their bodies more readily. Individual learned responses also helped our ancestors cope with changing temperatures. Migratory habits, such as moving to higher, cooler ground in summer and lower, warmer places in winter, helped them cope with the changing temperatures that coincided with glacial expansions and contractions. Hiding in caves was another way our ancestors learned to live more safely and comfortably, keeping their bodies in places with a more reliable range of temperatures.

It's no coincidence that the first creative art has been found in caves, by the way, because temperatures in caves are quite constant and moderate. Caves were relatively relaxed environments to our ancestors—places where cognitive learning, social organizing, and art had a greater chance of occurring.

Our ancestors also learned to control fire as a response to changing temperatures. It is not clear whether this was a social or an individual adaptation, but it appears to have found its way around the world quite early in our species' evolution (Gowlett, 2016). This suggests that controlling fire was not learned separately by different groups but was passed between groups after it was first accomplished. We will never know, but our later, widely demonstrated use of social speciation certainly suggests the latter explanation—that it was passed around from group to group (Diamond, 1997).

Being able to keep warmth or coolness within a certain, comfortable range has certainly been a motivated activity going back a long way. As we have already seen, humans have created this kind of environmental stability by organizing socially to build shelters, invent and spread the use of heating and air conditioning, and other "hacks" that deliberately alter the environment to reduce its otherwise widely fluctuating demands. This is why there are now permanent settlements in the subarctic, deserts, and other inhospitable environments.

Again, by altering the world to meet individual needs, we have created the conditions for further, more rapidly occurring hacks. Lower risk increases creation, discovery, and innovation, so, having established steady food and water supplies, kept extreme cold and heat out of our local environments, and otherwise reduced environmental stressors, more time is available to develop other hacks. In many places today, people are solving problems that weren't even recognized as problems 100 years ago and doing so in ways that weren't even considered possible in those same

100 years. We have so moderated the environmental pressures that used to be met by genetic selection and learning that we have accelerated hacking.

Now we have a new problem: How do we adapt to the hacks themselves? We're still adapting, but now our adaptations are at the global (rather than individual or niche) level, in response to the pressures constructed by our species rather than individual pressures related to nonconstructed environments. So, what are we adapting to in this type of second-order hacking?

SOCIAL IDENTITIES AND SOCIAL MARGINS IN A SINGLE SOCIETY

A good place to start trying to answer this question is with a simple observation. Modern societies have built and inherited many social species. In addition to family and tribe, social forms include everything from sports teams to megachurches, government bureaucracies to online businesses, and many more. Ancient family and tribal societies have somehow provided the seeds for an incredibly rich array of social organizations.

Applied psychology has begun to deal with some of the problems associated with this second-order social speciation. For example, it has become necessary for most people to take on multiple, sometimes quite different identities, depending on the social groups in which we are operating. This can be seen in everyday life as people wake up in a family home, commute to work, visit friends after work, then head back home again. So, for example, my role as a husband and father requires me to be empathic, take on tasks with little or no immediate reward, listen attentively to others' needs, manage conflicts, and express affection and commitment to my family members. When I step out of my home in the morning, I become a car commuter in a city, single-mindedly attending to the sociophysical world around me, anticipating hazards, behaving more or less aggressively or passively depending on the timing of relationships with other vehicles, and using my horn (and occasionally hands) to communicate simple motivational messages. Never mind the workplace, where the number and types of behaviors and their potential consequences border on the infinite—but the end results are still a handful of outcomes that can be important to self, family, and other social groups.

The outcomes of performing these many, often very different individual roles are many and various, as well. Research on work–family conflict deals to some extent with the motivational differences among these roles (French et al., 2018; Sanchez-Burks et al., 2015). Certainly, conflict among roles has been shown to lead to some unfortunate consequences for people who don't cope well with them. Poor job performance, family conflicts, work dissatisfaction, and psychological burnout have all been associated with various aspects of this general type of conflict (Nohe et al., 2015).

Recent research in this area has tried to reframe this as work–family "balance," focusing attention on ways to cope well with the inevitable strains that occur between our social roles (French & Shockley, 2020; Montani et al., 2020). Although family support and individual coping appear to have greater effects on successful outcomes (French & Shockley, 2020), only a few coping methods have been examined, most of which engage the same sorts of methods that lead to creativity and effective social organizing. For example, ongoing feedback during coping, managing initial role perceptions (Smith et al., 2013), and developing and asserting solutions (French & Shockley, 2020) may help expand possible ways to harness social organizing for managing social identity conflict. Although organizational efforts to support balance appear to be less effective (French & Shockley, 2020), there are almost certainly creative organizational solutions that have not yet been examined.

I frame these social identity conflicts a bit differently, taking an approach that is not focused solely on the work–family margin. I refer to the effects of *social margins* on our individual thinking and behavior more generally. Trying to manage the different expectations of intersecting social environments is one of the important challenges of modern life. There are so many social environments, and sometimes such different behavioral expectations, that managing the margins between them can seem bewildering. But developing social organizing strategies that satisfy the people in different social settings can also be transformative, providing guidance for managing social speciation in the interest of sustainability. For example, when my children were in the car with me in the city, my spouse and I would let them know in advance that "what Daddy says while he's driving is very different from how he usually acts. It's because city driving is very different." We even gave a name to this very different Daddy: "Daddy's a Chicago Driver." The transition between their expectations of my behavior in our safe, comfortable home and my behavior while driving risky city roads deserved explanation—and a separable social identity.

Understanding how to manage the expectations at the margin between social identities—even family and public road behaviors—is an important object of study. Approaching things in this way helps to clarify the essential nature of social margins for our evolutionary hack. This is because the conflicting motives at social margins are the breeding grounds for new social species. Although they do not always involve open conflict between people, they often create internal conflicts that elicit individual changes to help us cope.

Individual Adaptation to Social Margins

Going back to the temperature adaptation example, for most of human history, it was enough to learn new, individual behaviors. It was not until the evolutionary hack of organizing groups to deliberately alter the environment that a new type of adaptation became necessary: adapting to our social environment. Directing the navigation of social environments is an essential component of managing the social speciation process. Social pressures themselves pose powerful "environmental" constraints. They can create the conditions for speciation—or make it more difficult. We are deeply motivated to adapt to the social environments around us because this kind of adaptation allows us to manage broader environmental pressures.

If you doubt that we adapt to our social environment, think again about everyday life for most living humans. All of us have learned to cope with social margins, some more successfully than others. There are those who try to stay "true to themselves" and behave much the same in every situation, regardless of its demands. I used to refer to these as *characters*—people whose social motives are so constant that their behaviors are predictable regardless of their social situation. Others, such as career diplomats, pride themselves on understanding all the subtleties of the many social environments where they do their work. They have learned to change the observable part of their identity depending on their social situation. They are like actors in theaters, showing different social identities to different groups of people (Rafaeli & Sutton, 1987).

Researchers refer to this changing of visible identity as *emotional labor*. But it is not just labor in the sense that actors and career diplomats are paid for doing it. It is also labor because they must integrate conflicting identities. For example, after leaving work as a social worker—a difficult type of emotional labor because it sometimes requires hiding anger—some people return to their homes and rather than talking to their spouses about how angry they are at the bureaucratic barriers they encountered that day, try to express joy at their child's accomplishment in school that day. Angry feelings from work are subordinated to the need to show joy in the family.

This example points to situations—such as family—where the demands of the social identity are so central to our motives and so much a part of our daily habits that we aren't "actors"—we don't have to "fake it." When we enact the surface behaviors of such a social identity, it is called *deep acting* (Brotheridge & Grandey, 2002). Some people might also call this "being yourself." But the social worker example illustrates that it is not always easy even to be yourself.

How do we learn these social roles? The short answer is that we observe and try out different behaviors to test their outcomes. In other words, we learn social contingencies, just as we decipher contingencies in any kind of learning. For example, we learn that smiling at a child often leads to easy interactions rather than causing strife with the child's family. Learning these contingencies begins early in life, probably because it is so important to our individual adaptation—never mind our social hacking later in life.

Because these social roles are often important to us (Jones et al., 2003), people have powerful motivations to learn social contingencies and the behaviors required for successful social navigation. Using the temperature example, you may be fairly well suited to cool, subarctic climates because of your ancestry and may even have means for keeping cool on hot days, such as jumping into a cool body of water. These don't always rely on other humans. But wouldn't you rather deal with this important body temperature problem by stepping into a friend's air-conditioned house for a minute? Rather than driving to the water, taking off clothes, putting swimming clothes on, and so on, you can step into air conditioning for a few minutes, cool off, and be ready to go onto doing something other than keeping cool.

The convenience of a locally constructed niche far outweighs the advantages of other options when time is limited and other options are scarce. Talking a stranger into letting you enter their air-conditioned home might require a lot more emotional labor and is more likely to fail than having a friend nearby who will share their cooling quite readily. This is one example of how the management of social margins can keep us cool. There are many more.

But there is a more complicated answer, as well. Think about a situation where you really must have something—it is so important to you that if you don't get (or avoid) this outcome, you think your life will be over. For most, the love of another special person is this sort of motivator. Trying to enact the role of "mate" is a profound social motive, and it includes a key ingredient for a special type of cognitive change. Individual "learning" is inadequate for describing this kind of change at social margins. When social circumstances lead to fundamental changes in individual identity, it is often described as *developmental* change.

The Key Ingredient: Adapting to Conflict

This "wannabe mate" identity sets up some difficult expectations and illustrates this kind of developmental change at a social margin. Those who would become a mate need to show great, enduring interest and affection for another person, sometimes for an entire lifetime. Yet, once they become a mate, they may also be expected to engender a sense of security and nurturing for children resulting from this relationship. These two roles—committed, passionate mate and safe, nurturing parent—meet when the first child joins a family. And data are clear that this is the most common time for marital satisfaction to plummet in both partners (Twenge et al., 2003). Incompatible social expectations end up competing with one another, as when a nursing mother no longer has much time to show affection for her mate or a "breadwinner" role leads to the kind of anger-to-joy pivot described earlier as part of emotional labor.

Competing motivations lead to conflict—conflict that needs to be managed if the relationships involved are to survive and thrive. Applied psychology has shown once again that social support is important to this adjustment of roles (Berg et al., 2016; French et al., 2018). So, a helpful grandparent might come to the rescue for "date nights," during which successful new parents work out their expectations. As with any form of learning, this kind of social support can be essential to success. But it is also introducing a different social form to the family.

Just learning to clarify social expectations may be enough for some to manage this change, but for others, rearrangement of social circumstances may require developmental change. Again, social support may mean incorporating another care provider into the mate relationship—the helpful grandparent or a paid caregiver joins the family. This adaptation redefines not just the mate and parent relationships but also adds a new social identity to the "family" organization. This kind of redrawing of the map of the social organization often involves developmental change in one or more parties to the process. Thus, the developmental change that often accompanies social speciation is a motivated response to the demands of a social environment.

DEVELOPMENTAL CHANGE

The by-product of interpersonal conflicts like these is sometimes a meaningful change in perspective. Instead of seeing the social world using one's old map, enough of the right kind of conflict can force us to redraw our maps of the social world. Developing a new social identity means that a new cognitive map replaces the old map of social contingencies in an environment (Cruwys et al., 2014; Tedeschi, 1999). This kind of change is an essential outcome of many applied psychological interventions, including clinical treatment, managerial development, and broad organizational development (Jones, 2015, pp. 172–175). Understanding developmental change is likely to prove helpful for managing the social speciation process.

But developmental change is quite complex (Halford & McCredden, 1998). Perhaps the simplest way to describe it is a change in general perspective on one's circumstances (mostly social circumstances in adults). Instead of ordinary learning, where we draw new lines in an existing cognitive map, developmental change defines new contingencies between causes and effects. Developmental change forces us to throw away part or all of the old map used to navigate social expectations and try to replace it with a map that better describes the realities of our social environment. Social development also involves a fundamental motivational change caused by a conflict—loving and caring for a new family member in our example.

Developmental change can also be defined using the limiting factors that distinguish it from other kinds of cognitive change (Crain, 2010). First, it is not the same as learning. A different analogy than maps is that most people have had experiences in their lives where, instead of responding to something new by nodding and saying, "Yes, that fits my understanding," we have had to stop and say, "Wait, that's not right!" These are events where we observe profound conflict with our prior understanding, so much so that we are at a loss to explain them. Ultimately, we realize or comprehend something rather than just recalling and reciting its features. Developmental change is more like this latter comprehension type of cognitive change, where we put pieces of the world together using principles that are new to us (Halford & McCredden, 1998; Persons, 1993).

This leads to the second condition: Developmental change is associated with a more complex understanding of the complex world in which we operate (Jones, 2020, pp. 173–174). Some of the best advice I received when starting my first job was to "ask a lot of questions." This was good advice for several reasons. First, when I started asking questions, it made clear to others that I wanted to do well, I was willing to learn, and I acknowledged that they were more knowledgeable about the situation than I was. But it also meant that I actually did learn about the complexities of the new social situations I had entered. Rather than simply watching others and copying what they did in a rote fashion, I was busy building an understanding of the "rules" underlying these behaviors. I was building a more complete and somewhat more complex understanding of the expectations of the job. This led to considerably greater success than simply "copying," which is analogous to memorizing answers to a test. I was able to pass the test because I understood how things "worked." When my map was more or less complete, there were few tasks I was unable to understand because, even though I may never have done them, I understood the broader expectations and priorities of that particular workplace.

To elaborate: My first payroll job was as a busser in a country club. I learned that all food being served needed to be handled with clean hands and not be exposed to unclean surfaces (e.g., the floor). Dirty plates and dishes were to be removed quickly with at least nodded consent from the member affected, and trays were to be carried on a shoulder rather than in hands in front of yourself because this was safest and least likely to result in an accident. Most important, I learned to always be aware of members' needs and respond politely when a member addressed me. The list of specific work behaviors was not a long one, but it still helped to understand that the safety, health, and pleasant sensory experiences of the members were broader principles that guided the work. It was not hard to figure out how these principles could be applied in similar situations, and I was promoted to cabana boy for the summer months. (The movie *Caddyshack* [Ramis, 1980], which came out soon after, certainly resonated with my experiences.)

What was happening here was that the rote behaviors I observed in this job could nicely fit under *abstracted principles*—rules that combined groups of similar behaviors. This "map," comprising principles underlying social cause–effect contingencies, fit the complex social world of a country club. Similar maps are developed for other social and physical worlds in which we are trying to survive and thrive (Persons, 1993).

The third condition that defines developmental change, following from this understanding of principles, is that people who have experienced developmental change understand things in this different way, but people who have not put things together in this more complex fashion simply "don't get it" (Gabenesch & Hunt, 1971). Thus, new bussers simply tried to copy what others did before they understood why we did things the way we did. They might carry a tray in front of them when no one was looking, not understanding that this is a matter of safety, or smile but not respond to a request, not realizing that there was a need they were supposed to help meet beyond just being polite. The broader principles hadn't been developed yet.

Principled, abstracted thinking is one of the hallmarks of cognitive developmental change. People who have not engaged in this kind of thinking do not understand the world in the same way as those who have engaged in it. Before one understands principles guiding behavioral contingencies in a situation, one is guided by a completely different kind of contingency map. Every little behavior has some apparently unique set of consequences. Being able to recognize the similarities and differences between situations makes it possible to apply broad principles to local choices (Jones, 2020, p. 172). Not understanding these principles can be quite a bit less successful.

A Scientist-Practitioner Example of Individual Developmental Change

One of the most memorable conversations I have had was with Milt Hakel, arguably one of the most influential I/O psychologists alive today. It wasn't just having a chance to talk with so prominent and agreeable a colleague that made this conversation memorable. We were talking about the recent evaluation of a developmental assessment center that he had helped to set up over a decade before. Almost everyone who went through this corporate assessment center was reporting that it had made a big difference in their lives, but the long-term data from the evaluation study were not reflecting this kind of "developmental" change.

To understand what "development" meant here, we need to be more specific about assessment centers. These days, assessment centers are used by organizations all over the world for hiring, promotion, training, and development, mostly for high-stakes jobs, such as executives, managers, and security personnel. Originally, they were invented to select spies during World War II, which provides some good illustrations of how these centers work. First, they rely on job experts, who observe job candidates react to several realistic job scenarios. Knowledgeable psychologists are included as assessors, which is not surprising given the kinds of stressors in jobs like these. In one of the exercises in World War II centers, the spy candidates would be given some background information, then have a short time to arrive at a plan for "gathering information in an occupied territory" with a group of other candidates. The candidates in this leaderless group discussion would present their plan to the assessors, who would then recommend which candidates to assign to this dangerous duty. In developmental assessment centers, the job experts take things a step further, providing recommendations for candidates to follow to prepare them for future jobs (Thornton & Rupp, 2006).

Assessment centers can be intensive experiences, not just because of the high-stakes jobs they are used for but also because candidates face scrutiny from people high in the organizational hierarchy who will make important decisions about the candidates' future. Not surprisingly, candidates often report that they come away from centers with very different understandings not only of the jobs but also of themselves and their relationships with other people (Thornton & Rupp, 2006).

Back to my conversation with Milt Hakel: The evaluation study had turned up all kinds of good news about outcomes of importance to the sponsoring organization. Assessors' (secret, independent) evaluations were predictive of candidates' later performance, as well as promotion into the upper management jobs they were being assessed for. Despite some conflicting feelings immediately after getting the feedback, candidates reported high satisfaction with the developmental feedback they had gotten, and even assessors had seen the center as a learning experience—for themselves.

But the difference that mattered most—whether people who had gone through this powerful experience were more likely to find their ways into executive jobs—wasn't showing up in the data. Because of scheduling problems, not every candidate nominated for assessment was able to go through one of the two to three centers scheduled each year. It turned out that people who hadn't been assessed were as likely to advance in the organization as those who had gone through the experience. From the point of view of organizational decision makers, the "developmental" reason for this expensive assessment center was not being realized.

Milt's question was this: Candidates reported developmental life change from their experiences in the center. So why didn't this show up in the outcomes that mattered to the organization? Even though participants thought that the experience triggered profound change, why weren't the data showing this? To answer this, let's assume for a moment that all the assessees were wrong. What they saw as a powerful experience hadn't changed how they thought or behaved moving forward. The study results shed the light of "reality" on their biased views of their experiences. Certainly, what we know so far about the long list of heuristics and biases demonstrated in psychology would support this view.

But what happens when we assume the opposite—that, in fact, the evaluation study didn't capture "reality" fully. The assessees were right about the changes they had experienced, but, for methodological reasons, the evaluation itself had missed key aspects of these experiences. Looking more closely at a couple of issues in developmental psychology, as well as glancing at the progress of psychological research methods, suggest that this is the better answer.

Theory and research on developmental psychology strongly suggest that, unlike behavior change and cognitive learning, Milt was hinting at some of the problems psychologists have trying to "capture" developmental change. Remember that developmental change is defined as a reorganizing of contingencies using abstracted principles. If the evaluation had measured individual behavior change—not changes in the principled reasoning of assessees—it was not likely to capture this kind of (developmental) change.

Another of the conditions of developmental change—that people who have not experienced it don't recognize when it's happened—may also help explain why it wasn't captured. Specifically, the evaluation researchers had not been through this same kind of experience and were therefore asking the wrong questions.

Finally, going back to the first condition describing developmental change: It is not the same as learning. Most of us intuitively recognize the difference between remembering something when we're asked to and realizing something. Many of us remember the experiences of our first sexual encounter. But, for many of us, we also realized something fundamental about the nature of adulthood—that people fall in love and get married more because we sort of "have to" than because we are making clear-headed decisions about who we want to hang out with.

Getting back to Milt's question, let's assume for a moment that a good many of the assessees had experienced such realizations as a result of their assessment center experiences. Their thoughts about what it was like to work a managerial job had been fundamentally altered to the point where they were no longer working with a sort of simplistic "map" but an entirely new way of seeing the activities they would be engaged in should they be promoted. Their reports of having experienced fundamental change were real: The old maps they had been using to navigate their professional lives were replaced by more accurate, complex descriptions of the contingencies of the situation. In other words, they had experienced developmental change. It's just hard to identify this kind of change.

How Adult Developmental Change Gets Missed

I like to use a personal experience to illustrate this. Before our first child was born, I asked a few close friends and family about what it had been like to become a parent. Everyone gave the same answer at first: It was indescribable to those who hadn't experienced it. Every answer I got was a close variant of "You can't understand it until you've experienced it." I pressed the point, wanting to know more but also with the belief that, as a graduate student in psychology, I could understand the experience.

From this point on, everyone had different, deeply felt descriptions of different points in the experience—the realization before childbirth that they were now going to be responsible for another's life, the bloody process of childbirth, the engrossment and bonding with an infant that follow immediately after birth, and the change in priorities that occurred in the first weeks of the new child's life. I nodded in agreement. Being a student of psychology, I now understood. But, of course, I didn't until my first child was born.

This variety of answers provides one more window into the nature of developmental change. First, and again, it is profound. Having a child is a defining life event for many people, on par with falling in love, getting a degree, losing a parent. It redraws our maps of the social world. Second, this type of change tends to occur differently for different people at different points during triggering life events. It is defined using different words, alludes to different aspects of the experience, and even has somewhat different repercussions. In short, it is idiosyncratic, experienced quite differently by different people. Third, and once again, developmental change is a complex kind of change. If it's hard to understand "unless you were there," there is probably a lot of explaining involved in trying to understand it.

But this first answer—that you can't understand it until you've experienced it—is a defining description of developmental change. This kind of realization can now be understood as a qualitative change in cognitive maps. The early developmental psychologist Jean Piaget labeled this kind of change *accommodation*, distinguishing it from the more common cognitive change *assimilation*. He argued that assimilation occurs as we add new information to existing cognitive maps (which he called *schema*), with accommodation as a fundamental restructuring of our schemas. Assimilation occurs often and is incremental, while accommodation occurs only occasionally and "all at once" (Lourenço, 2012).

Recent scholarship in adult development seems to have left the "sudden change" (stage) notion aside. But it also may be the result of a recurring problem related to Milt's question: It is difficult to uncover this kind of sudden, idiosyncratic, complex change using current measurements and methods. The answer to Milt's question is that current methods cannot pick up developmental change. People who'd been through assessment centers could articulate that profound change had occurred, but the nature of this change was quite different for different people, such that the questions asked in the evaluation study survey were not able to capture them.

Think of it this way: Electrons have existed as a phenomenon basically forever. We only got the idea of an invisible particle that is the basis for all matter in about 350 BCE, when Aristotle suggested such a thing. Worse, we were only able to measure the activities of electrons a couple of centuries ago with the harnessing of "electronicity" and got a look at an actual atom just last century through the wonders of electron microscopy. How hard do you think it will be to measure developmental change in a similar precise fashion—never mind even being able to identify it scientifically with the measurement instruments we invented just last century?

Another way to understand the problem of identifying developmental change is by asking people about their experiences in college. Colleges claim that they are a place where developmental change can occur, and people spend huge sums to try to experience some kind of profound change. This is consistent with the experiences of many people who went through college, so there is some agreement up to this point (Davies, 2011; Kuhn et al., 2000; Songer et al., 2009). But try demonstrating this to funding agencies or to first-generation family members who are being asked to spend a small fortune on college education. Those of us who've experienced development as a result of our college experiences have a big problem trying to explain what happened during the several years of discovery that may have culminated in a realization one night after a drink with friends that the world looked very different all of a sudden. As a family member who has never been to college, why wouldn't I just opt for the "drink with friends" as a moment to aspire to, especially if I'm inclined to avoid the risk of spending tens of thousands of dollars on an uncertain outcome? Those of us who have "been there and done that" know it's there—we're just wrestling with a measurement problem. Just because we can't measure developmental change doesn't mean it's not happening.

Development of Long-Term Perspectives

Developmental differences in time perception may be especially important to people's perspective on sustainability issues, in particular (Jones et al., 2016). The rush of modern life relies on a perception of time as a scarce, immediately consumable resource (Cialdini, 1988). By itself, this creates some conflicts. Many kinds of work illustrate how this urgency creates a tradeoff between speed and quality (C. Hedge et al., 2018) and may lessen the kind of calm that can encourage learning. Recent data also suggest that developmental change proceeds toward a long-term perspective, which may be especially important to sustainability (Baltes et al., 2014; M. Fisher et al., 2019) and creativity (B. Koh & Leung, 2019). This is where people adjust thinking and behavior in light of the internal conflict arising from the recognition of our mortality—integrating the fact that "life is short" into our thinking and actions (Jones, 2018). Interestingly, Jeff Bezos of Amazon thinks long-term thinking is the most important ingredient for success (Umoh, 2017).

Even more interesting is that there is a bit of evidence that long-term strategic perspective can be changed through certain kinds of life experiences (M. Fisher et al., 2019). This could help to deliberately manage the developmental change involved in social speciation processes. But first, there is a problem that needs to be solved. Because developmental change often happens idiosyncratically, it may be difficult to find experiences that work for a large group of very different people. This would make targeting an entire social species for developmental change challenging.

Exhibit 5.1 illustrates this. A group of highly successful, mostly older, mostly White people whose scores on a questionnaire identified them as later stage thinkers were asked to share the life experiences that had given them their long-term perspective (M. Fisher et al., 2019). These life experiences were then listed for a larger group of people to check off after completing the same questionnaire. Of the 30-some-odd experiences the larger group checked off, the 10 listed in Exhibit 5.1 were the ones that correlated significantly with the larger group's scores on long-term perspective. It is not obvious how these experiences might have a relationship with each other, and the data showed the same lack of relationship among them.

But it did suggest that there may be some places where development is more likely than others. For example, travel that entails exposure to a different society or culture is thought to incite such change. But this is only under circumstances where the traveler is open to the differences they see (Terry, 2020). Understanding how this happens may be a rich area for further inquiry.

> **Exhibit 5.1**
>
> **Top 10 Life Experiences Associated With Developmental Change in Time Perspective (In Order of Strength)**
>
> 1. Provided advice and direction to others regarding their future
> 2. Achieved a self-set health goal
> 3. Watched a documentary outside of personal interests to learn something new
> 4. Spent significant time talking and listening to an elderly person
> 5. Purposefully stepped into a difficult situation to experience new ways of learning and self-growth
> 6. Invested personal finances before the age of 30
> 7. Actively sought out information about one's own personality traits or strengths and weaknesses
> 8. Read a book about travel just for fun
> 9. Participated in a problem-solving and decision-making exercise as part of a job or other group membership
> 10. Listened on purpose to a podcast about a point of view different from my own

DEVELOPMENTAL CHANGE IS SOCIAL

How does developmental change affect our urge to alter the world to suit our needs? Along with the conditions for defining developmental change that we visited earlier—triggered by conflict, not the same as learning, complex, profound, and sometimes sudden—developmental change is social. The urge to manage the demands of the environment may be like nesting for birds—primarily for the safety and welfare of local social groups. It is not just facilitated by social support. Our evolutionary hack is socially motivated and requires social invention to realize. Exhibit 5.1 hints at this because seven of the 10 experiences have clear social components.

Psychological research supports this axiom profoundly. Perhaps the most obvious sources of support for fundamental social motivation come from discoveries about emotion. Research on emotive responses shows that like other social mammals (e.g., dogs, porpoises, chimps) and unlike less social ones (e.g., bears, wolverines, tigers), we have elaborate display systems devoted to expressions of love, anger, fear, sadness, and so on. In humans, these are mostly demonstrated in the most visible places on our bodies—our faces. We all express emotions, often many times in one day, mostly in the presence of other people. Most important, people around the world share a common understanding that smiling people are happy, frowning people are angry, and so on—across all cultures (Ekman, 1992). This and considerable other research (e.g., Feng et al., 2021) is evidence that engaging in social interaction is a fundamental human motive.

Recent research delved further into emotive displays (see Jones, 2020, pp. 144–148). The evidence largely supported this view but with some important complexities that tie back into the conflicts we experience in social margins. Facial displays of emotion are the basis for inferring (correctly or not) others' motives. Here's how this works: When you smile, I instantly infer that something is making you happy. You see something desirable approaching, which arouses you (and me, when I see the smile) to be ready to get that good thing. When you frown, I worry that you are angry or sad, that something you are reacting to may get in the way of the smooth operation of our lives in the moment. You might be seeing risk or missing something desirable, but either way, you signal to me that it's time to get moving.

This recent research has clarified some of the complexities of social motives, but we are only starting to understand how interpersonal relationships are influenced by emotive displays (Tran et al., 2011, 2012). Still, when the use of complex language is added to research about emotive displays, the evidence is overwhelming. Humans use all kinds of words and symbols across all known languages to efficiently communicate the particulars of our motive states.

Social Support for More Than "Just" Learning

The complex process of learning helps to explain some of this social aspect of development as well. One of the core concepts from learning is the idea of *transfer of training* or just plain *transfer*. Kevin Ford and his colleagues (Blume et al., 2010; Ford, 2021) have spent the last few decades trying to describe transfer and understand how and when it works. Like many scientific constructs, its definition has changed as we have learned more about it. But transfer is very much an applied concept, so its definition has remained pretty practical: Transfer has occurred when learning from one situation is demonstrated in another.

I/O psychology is concerned with how contingent relationships (A leads to B, and so on) learned in one environment (i.e., in a training course) are used to solve problems in another (i.e., the actual work environment). This is called *positive transfer*. One of the clearest insights from Ford et al.'s (Blume et al., 2010; Ford, 2021) research is this: Positive transfer occurs when people have social support in both the learning and application of the learning to a new environment (Blume et al., 2010; Huang et al., 2015; Hughes et al., 2020). Put succinctly, transfer relies heavily on social motives. To be included in and valued by some group is so important that we will risk trying something we've learned in one circumstance if we think it will serve our group in another circumstance. We figure things out together.

Being social provides for lower stress and longer, happier lives (Yang et al., 2016). This is reason enough to stay social. But social interactions also lead to conflicts, and under the right ("safe") conditions, these can lead to developmental change, too.

Social Support, Development, and the Paradox of Conflict

It is still possible that social speciation occurs as a result of "relaxed rat" cognitive learning. This is consistent with (a) the generally higher degree of individual comfort and security in many modern societies in the past 50 years or so, coinciding with (b) lots of new ideas and inventions, including invented social structures.

But there are a couple of big problems with relying on this correlation as a basis for future decision making. First, if cognitive learning is the basis for our hack, we are using our existing cognitive maps to make these innovations. But our problem for a sustainable future is that we need to control this very inclination—to use our existing approach to solve new problems, problems that are qualitatively different than keeping us comfortable as individuals. Second, the environmental problems we're addressing posit serious risk—we have already seen how heuristic decision processes take over from relaxed rat thinking when risk is foremost in our minds. Clearly, even if cognitive learning can help us to control our urge to use cognitive learning to make ourselves safer and more comfortable, we still need another kind of cognitive change that does not rely entirely on this kind of learning.

The good news is that developmental change has also played a role in the societal changes we have seen in recent decades. This is because developmental change is where we entirely "redraw" cognitive maps—sometimes sharing the new ways of thinking that develop into new shared mental models (Tasca, 2021). Developmental change is defined as "accommodation" of mental models to the discovery of a new reality or perspective on reality. Such developmental leaps are often motivated responses to conflict with others. Thus, the assumption here is that attempts to manage ecohacking will need to target and manage developmental change—and the conflict from which it arises.

More specifically, managing change in shared mental models relies on managing social interactions and especially managing conflict. We have seen that conflict often occurs at social margins, which is also where creative solutions are bred and new mental models formed (Tang, 2019).

But calm command of the principles ruling a situation is unlikely at a social margin, where two groups with different expectations are coming into contact. Emotions are displayed and encoded, different languages expressed, and motives communicated with greater or less accuracy in these margins. Finding a way to deal with this uncertainty and the conflict often following from it can lead to new perspectives, restructuring, and creative social solutions (Montani et al., 2020; Sanchez-Burks et al., 2015). Such conflict-motivated change in developmental perspective is key to the social inventions that are at the heart of our evolutionary hack.

SUMMARY

Understanding how to incite the peaceful conflict that sparks a change of perspective is the approach that will help to manage the social structures of our hack. Within a society, individual developmental change occurs at social identity margins. Between societies, new social species occur at social margins. We form new social groups to try to moderate the effects of our environment to make ourselves more comfortable and safer. Conflict makes us feel uncomfortable and unsafe. So, one other essential process needs to occur for social speciation to succeed: Conflict needs to be managed—we need to feel safe for development, creativity, and social speciation to occur. We turn to this next.

6

Making and Managing Social Ecotones

Depending on how effectively we mold our social organizations, we can hack our social environment. By moderating the extremes of social interactions, applied psychologists can help coordinate the activities necessary to build shelters, jets, art, and markets more sustainably.

In this chapter, I argue that the process of social speciation occurs most often in the social margins between groups, what biologists call *ecotones*, in reference to the margins between biological communities more generally (Presotto et al., 2018). The previous chapter described how conflicts in an individual's social margins increase the chance for individual developmental change. Here, I explore how mimicry, scarcity perceptions, and conflict in these margins—social ecotones—can lead to distinct social species, and I discuss some of the tools available to scientist–practitioners to help manage this speciation process.

https://doi.org/10.1037/0000296-006
Sustainable Solutions: The Climate Crisis and the Psychology of Social Action, by R. G. Jones
Copyright © 2022 by the American Psychological Association. All rights reserved.

SUSTAINABLE SOLUTIONS

YARN BOMBERS

Have you ever heard of "yarn bombers"? One common bit of advice to people contemplating retirement (a new type of social species itself) is to "keep busy," by which advice givers mean "find something to do," often with a group. Yarn bombers are this kind of group—sort of. They include elders looking for sociable fun in a time-honored activity: the knitting circle. But yarn bombers also include young adults who love arts, including knitting but also graffiti. Yarn bombers knit things that can be put onto objects in public spaces, such as stop sign poles, junction boxes, trees in the boulevard, and bollards. Yarn bombers are graffiti artists who use yarn instead of spray paint and are perhaps less fleet of foot and physically daring than the stereotypical graffiti artist.

Yarn bombers can be considered a new social species. This chapter examines the variables involved in forming and managing groups such as the yarn bombers to support a sustainable future—to adapt and survive. New social species have a few characteristics that make them particularly important to managing the human evolutionary hack—and these features are uniquely understandable and manageable using human psychology. First, they fill important, existing, often immediate human needs. They are not formed to create new needs—successful, sustainable social species seldom are. All kinds of species match some capacity that members of a species possess with the physical demands of a niche. Social species are similarly reliant on meeting the "needs" of a particular "niche." Marginalized young adults with a slightly rebellious, creative urge and isolated seniors who love productive social activity meet with the yarn bombers to produce ornaments in otherwise ordinary public spaces that make passersby smile—or at least not call the cops.

Second, because social species are hacking the usual niche adaptation process, they are often meeting the same general need—to moderate some of the extremes that occur in social environments. For example, the extreme isolation felt by some seniors, the social conflicts caused by frustrated youths who are not yet fully clear about the contingencies of their actions (getting arrested for graffiti), and the ugliness that often accompanies urban life are all these sorts of social extremes. The yarn bombers' activities serve to address and potentially mitigate these extremes.

Third, social speciation combines aspects of different types of groups that have proved successful before. In the yarn bomber case, there are elements of old-fashioned knitting circles, youth street gangs, local arts councils, and social action committees, among other preexisting social forms. Each of these social species has at one time or other fulfilled important existing needs.

Fourth, because of the relative safety and social support that have been building in many prosperous societies, new social species have grown dramatically during the past 100 years or so, following the violent conflicts that marred the latter part of the industrial revolution. Social species are formed where people see themselves as having the resources—time, energy, yarn, fast legs, broader social approbation—to support their social speciation. They can take advantage of current munificence to forestall extremes associated with their circumstances (isolation, conflict, crime, urban blight in the yarn bombers).

Finally, social speciation occurs as a result of developmental change in some (but usually not all) members of a social species. The characteristics of developmental change make it an especially difficult target for deliberate management, as we saw in the last chapter. It is triggered by conflict and is more complex, profound, and sudden than learning. But because developmental change is a response to social dissonance, it is a key ingredient for managing social speciation. At least one member of a group and perhaps members of the interacting groups need to have developed a perspective that precedes creative social organizing.

SOCIAL ECOTONES

As Jared Diamond (1997) argued, there are often strong pressures to adapt when two societies or cultures meet. This is also true at the level of smaller social groups, whether they are deliberately combined (as in yarn bombers) or (more often) occur naturally. Although organizational sciences have gained considerable expertise in the development of new social forms during the past few decades (Kirkman et al., 2000, 2016; Schommer et al., 2019), applications designed to manage social speciation for a more sustainable future have received less scrutiny (Gallagher et al., 2020; Mannen et al., 2012). To advance in this direction, we need to understand the development of social species in social ecotones.

Automatic Imitation and Mimicry

In some ways, altering behavior in response to others' behavior is like other adaptations. Genetic selection, migration, learning, and other responses to environmental pressures lead to changes in forms and behaviors (Snell-Rood & Steck, 2019) regardless of whether the "environment" in question is ecological or social. Mimicry is just another one of these reactive adaptations to social environments—at least partly. As in other species, automatic imitation occurs when we interact with others within our social group—no complex cognitive maps are required (Cracco, Bardi, et al., 2018).

This sort of reactive imitation may play some part in our evolutionary hack. Most of the large body of psychological research here deals with facial mimicry: Young children watch their caregivers' smiles, frowns, and surprise and copy these expressions at an early age (Cracco, Bardi, et al., 2018). Automatic imitation is thus an important form of rote learning of the social roles discussed in Chapter 5. By observing and trying different behaviors in different situations, then discerning their consequences, children learn social contingencies, just as we decipher contingencies in any kind of learning (Cracco, Bardi, et al., 2018).

But mimicry appears unrelated to automatic imitation (Genschow et al., 2017). Instead of the rote automatic imitation of infancy, mimicry has some of the hallmarks of cognitive learning—constructing abstract principles in a cognitive map. It tends to happen within social groups (Genschow et al., 2017) and is inclined to be prosocial rather than antagonistic (Cracco, Genschow, et al., 2018). In fact, mimicry occurs when groups try to coordinate their activities (Vicaria & Dickens, 2016). We may quite deliberately create copies of what others do—updating our social-cognitive maps. But this occurs more within social groups than between them. This is probably the kind of individual learning that defines social roles within a group, but which comes into conflict with expectations in social margins with other groups.

It may seem obvious, but television theater relies entirely on coordinated, often quite sophisticated mimicry. It seems obvious because actors are mimicking the social behaviors and identities that viewers expect to see in similar circumstances. Viewers are comparing their expectations about how to act (called *scripts*, not coincidentally) with the actions onscreen.

Watching such mimicry can be engrossing, probably because of the enormous importance of such socially coordinated mimicry to our hack. Only if viewers know the "usual" expectations of a social identity can they learn ways to deviate without causing unmanageable conflict. Instead of experiencing conflict directly, viewers experience the surprise that precedes conflict. Dramatic and comedic entertainments usually deal with conflict-laden topics—love and fidelity, jealousy and hate, inequity and justice, war and peace. We learn by watching others try to cope with these conflicts in a "safe" social setting like our own. As a result, we learn the "surprise" contingencies of social behavior in a social species something like our own. Given enough of a difference from the expectations of our scripts, however, we may even learn new, broad contingency rules. This is why theater can be "transformative" and can lead to developmental change.

Mimicry and Conflict at the Social Margins

Mimicry that relies on cognitive learning may be the basis for the new social organizing that has led to human alterations of the environment. But humans copy social groups other than our own, as well (Diamond, 1997; Morris, 2010). It is not a great leap to suggest that mimicry of other groups may be a starting point for the conflicts and developmental changes that lead to new shared mental models—and new social species.

On the broader stage of history, mimicry can be seen this way, functioning to moderate conflict on the margins of social identities. Lawrence of Arabia, Mohandas Gandhi, and Geronimo all took on the garb and manners of cultures other than their cultures of origin to try to manage the conflicts between these cultures. Conflicts had raged (respectively) between Islamic and Christian, Indian and British, Indigenous and European cultures for centuries. Lawrence, Gandhi, and Geronimo were high-profile members of one of these cultures, so their decision to visibly emulate the social identities of the "other" cultures was a surprise. If you remember that humor and creativity are examples of reframing, it should not come as a surprise that their deliberate mimicry had some positive results. Ultimately, the blending of cultures in these examples was followed by reforms and hybrid social forms.

Following the Indigenous European example illustrates how new forms emerged even in the presence of severe, violent conflict. Native Americans of the Great Plains started regularly interacting with Western Europeans when fur traders arrived in the High West in the 1830s. These two cultures shared enough social conventions to generally manage peaceful, even profitable trade. Perhaps the most famous peaceful interactions between people from these cultures occurred during the Rendezvous, where trappers and traders from these groups had days-long parties at a designated place and time (Ross, 1924). Here they experienced both the expected social contingencies and the surprises that each culture had to offer—in a safer environment than the one they had been operating in for the months prior.

Not long after these joyous celebrations ended, violent conflicts broke out between these same cultures. Western Europeans of the time had developed their economies around the social invention of individual land ownership. Many Northern Plains cultures viewed their relationship with the land as a sacred trust rather than an "ownership." The misunderstandings that arose relating to this foundational difference in social species led to bloody conflict but also the eventual adoption of important social inventions. Perhaps the most notable of these is national ownership of enormous tracts of land in the Western United States—something that had not been seen before in recent European history but was adopted by President Theodore Roosevelt, who had many interactions with Plains people.

It is hard to determine whether Roosevelt was relying on social mimicry—and the new perspective he had gained from his sojourn in the frontier West (Morris, 1979)—but it is certainly a parsimonious explanation. And there are many other examples of social mimicry at the margins of societies, where specific behavioral changes—in clothing, food cultivation, languages, social relationships, and basic framing of events—arise when behaviors from one social species reform another. Clothing fashions are obvious examples of this sometimes, including the adoption of one culture's dress by another. The point is that new social forms happen naturally and sometimes through a sort of deliberate mimicry—even under conditions of considerable conflict.

ECOTONES AS VENUES FOR SPECIATION

Ecological science has identified ecotones as environments from which new species evolve. This process of speciation is thought to occur most often around these margins between two or more ecological systems (Cooke et al., 2014). A common example of a biological ecotone is a shoreline, where species that live in water and species that live on land and in the sky are forced together by the forces at work in their environments. These can include pressures in their main habitat, such as increased population and lack of available nutrients, and larger forces, such as tides and winds. In geological time, these larger forces made entire portions of the planet into ecotones, as when a comet impact ended the Cretaceous period (the age of dinosaurs) and forced remnant populations of a wide variety of species to interact. This marked the beginning of the dominance of birds and mammals—species that had evolved structures that allowed them to adapt to dramatic temperature change, increased terrestrial competition, and other changes that the comet impact brought.

Our own species is thought to have evolved at the ecotone between forests and grassy savannahs in sub-Saharan Africa. Unlike most of our primate cousins, we are not able to live comfortably in trees without the help of treehouses. Many other primates can climb and swing from trees with great agility and make nests and even sleep comfortably in a different tree every night. Unlike these primate kin, we can grasp objects with opposable thumbs and walk almost completely upright, abilities that suit us better for survival in open grasslands—mostly due to evolutionary changes brought on by dwindling forests and a growing ecotone (Jablonski & Chaplin, 1992).

Resources may be more available in ecotones, combining the resources of the multiple ecosystems that touch them (cf. Abwe et al., 2019). Paradoxically, survival in an ecotone may be more hazardous than survival in a niche (Hamm & Drossel, 2017), at least when species rely heavily on genetic transformation. Species that have adapted into a narrow, isolated niche tend not to do well in ecotones. They have not retained the requisite variety of genetic code necessary to adapt quickly to the many varied demands presented in an ecotone (Case et al., 2005). Species such as pandas and gorillas, for example, are supremely adapted to their particular forest habitats but have not fared well in the wider world. Their genomes have narrowed over time to put more energy into specific tasks. However, species, such as cockroaches and pigeons, that have been lucky enough to retain some advantageous variations can thrive in an ecotone by changing forms to meet its demands.

When we start talking about adaptation through learning new behaviors, the effects of ecotones on speciation are less reliable (Snell-Rood & Steck, 2019). Animals such as seagulls and crabs make their livings cleaning up the available nutrients in the ocean–shore ecotone. Seagulls are quick learners. Crabs are not. Yet both have done well. Likewise, some of the animals with cortices have done extremely well in the ecotones created by human changes to the environment (e.g., rats), and others have practically gone extinct (e.g., elephants and certain whales). Of course, human predation and habitat destruction explain some of these extinctions, but human attempts to wipe out rats remain unsuccessful. So, the kind of learning made possible by the cerebral cortex is not as clear an explanation as it might seem (Harari, 2014).

What is clear is that new species often emerge from ecotones. Species that have evolved to exploit the particular resources in their niche (e.g., pandas eat bamboo all day) and protect themselves from its threats (pandas are big and strong enough to ward off most would-be predators) are not likely to thrive where new, diverse resources and threats are found—as in ecotones (where there is no bamboo). Species that, by gene pool luck or behavioral learning, can adapt to the threats found in ecotones are more likely to find ways to adapt. This adaptation can take many forms but generally involves relationships with other species. These might include relying on other species in the ecotone, hiding from them, or dominating them. But both original species tend to speciate in the process (Roberts, 2017).

Our own species is a prime example. Standing on our hind legs, grasping tools more firmly with a hyperflexible thumb, developing longer childhoods, growing bigger brains, and other genetic changes are thought to have occurred partly in response to changes in the savannah–forest ecotone (Jablonski & Chaplin, 1992). These changes helped our ancestors to find resources and avoid dangers more successfully. And, of course, humans also found ways to hack the ecotone using cognitive maps, curiosity, and social organizations.

As per our usual sidestepping of evolutionary pressures, humans continue to try to manage the social environment, developing novel social forms and processes that make it possible to avoid the extremes of these environments, as well.

Social Speciation in the Ecotone

Human social groups speciate in social ecotones. We construct social species in the places where we must rely on interactions with people from groups other than our own (Dokko et al., 2014; Tang, 2019). Just as conflicts at the margins of individual identities lead to developmental change, so creative social speciation occurs in the margins where social identities meet (Pettigrew & Tropp, 2008). Inventing a new species to forestall conflict and scarcity is more or less the social equivalent of individual development. Although individual developmental change is a key driver of creative perspective, it is through managed conflict, perceived munificence, and surprise at social margins that disparate groups get creative (Dokko et al., 2014; Sanchez-Burks et al., 2015), occasionally solving problems that seemed insurmountable before these groups solved them. The process and motivations for social speciation and individual development are similar (Pettigrew & Tropp, 2008; Yong et al., 2020).

Examples of flourishing new social forms reflect a combination of these factors: cultural mimicry, prosperity, and conflict management. Historical examples, including late Mongol-ruled Bagdad, several 15th-century Italian city-states, 16th-century Antwerp and Amsterdam, 19th- and 20th-century Hong Kong, 20th-century America, and 21st-century Singapore, all demonstrate these conditions on the national scale, blending different societies in prosperous places, all of them seeking safety and comfort. Some other historical examples lack one or more of these three elements, such as the horse culture that emerged in Western North America after the arrival of Spanish horses in the 16th century. Native American tribes adapted some new internal social habits but faced intractable conflict from the often-desperate European groups expanding into Native Americans' tribal lands. Similarly, the trade routes through the Sahara Desert in the 19th century were able to thrive as long as they followed existing Islamic laws that moderated the extremes of social behaviors. Having not adopted new social organizing methods eventually doomed these trade routes to scarcity and conflict.

On a smaller scale, there are many rich examples of disparate groups coming together to solve some mutually important problem. The Manhattan Project is an interesting example, where scientists, engineers, mathematicians, highly skilled tradespeople, and expert social organizers developed and built the horrifyingly powerful bombs that ended World War II. Say what you will about the destructive—versus creative—outcome of this group's work, they organized themselves in novel ways that solved a previously unsolved problem. Theirs is a remarkable example of social speciation through the deliberate creation of a social ecotone, blending the perspectives and abilities of a large variety of social groups to forestall the most devastating of all known anthropogenic extremes—nuclear war.

Scanning the broader modern social world illustrates the importance of ecotones, as well. Some of the most outstanding instances of new social organizations today have grown up in some of the most culturally diverse places, such as Singapore, California's Bay Area, and other prosperous metropolises (Choain & Malzy, 2019). Again, these are places where most people are not struggling to survive and where a large portion of the population is comfortable enough to test their curiosity if they are so inclined. These curiosity tests are often deliberate mimicry of social forms and fashions. As urbanist Richard Florida (2005) famously pointed out, these are also places where many languages are spoken, many cultural backgrounds and experiences are celebrated, and people generally find ways to get along together despite these sometimes wide differences.

The common needs to develop safe, comfortable environments—and avoid the nastiest consequences of social interaction—have led to the kind of managed conflict, developmental change, and creativity that drive our hack.

Forced Proximity Amidst Prosperity

An obvious consequence of the tremendous success of our evolutionary hack has been the exponential growth of human populations. What is less obvious, perhaps, is that we have evolved far more dense blends of cultures. Many argue that such interactions between cultures lead almost invariably to warfare (Diamond, 1997). But the opposite seems to be the case more recently. There has been a major drop in the number and proportion of people dying in armed conflict in the years since World War II (Rosling et al., 2018, p. 114; Szayna et al., 2017). Instead, the density of hybrid social species appears to have increased (Crook et al., 2013; M. Lee et al., 2020; Quelin et al., 2019) as growing numbers of humans have come into greater contact, usually in urban settings.

What has changed? How have we avoided massive, destructive conflict and devised new social forms instead? One answer is that when there is enough to go around, and people are unable to completely avoid each other, we are increasingly motivated to manage conflict peacefully. Motivationally, more of us know what it feels like to have a comfortable, safe life—and would prefer to keep it that way. Why would we resort to physical conflict in these circumstances?

Violent conflict still happens in prosperous societies, of course, but it seems to be limited mostly to the groups in society who feel desperate—stuck in seemingly intractable circumstances that keep them from feeling comfortable and safe. If life is uncomfortable enough, for whatever reason, there is a point where one will choose to put it at risk. A tragic example is someone whose anxiety, anger, depression, or guilt leads them to attempt suicide. But for most people today, finding other ways to resolve conflicts—within or outside of us—is both desirable and practicable.

Recall that developmental change is defined largely by a new perspective on existing issues. We change the perspective from which we are looking at a problem or opportunity, altering our social identity to fit new circumstances. Prosperous social ecotones, with processes that moderate conflict among social identities, are key to both developmental change and devising new social organizations. Understanding this reality opens many ways to manage social ecotones, especially because it doesn't rely on developmental change in all members of a group. Having enough group members with perspective on the conflicts and scarcities of the ecotone they confront may be enough to manage creative social speciation. Developing a shared mental model only requires cognitive learning—not development.

Managing Mimicry, Scarcity, and Conflict

Scientist–practitioners have many tools to steer the speciation process toward a more sustainable future. Managing mimicry has not been an obvious part of psychologists' repertoire, though many of the tools commonly used to support decision making and change are based on successful previous approaches—a form of mimicry. A few organizational scientists have addressed this gap (Brauer & Wiersema, 2012), showing that the timing of mimicry may matter. But social speciation is commonplace in the business world, despite the lack of research evaluating it. Here, narrowly defined profit motives lead to a large number of new businesses, often by mimicking other businesses. These arise at the meeting of profit-seeking social motives (entrepreneurs) and people seeking to serve all sorts of other motives. These ecotones have bred enduring business forms for millennia. And it's easy to identify the kinds of businesses that serve essential human motives—grocers provide food, building trades provide housing, banks provide security, and less socially sanctioned organizations provide for human addictions (drugs and alcohol, sex, gambling).

Yet, many new businesses fail during their first years of existence. Those that survive often have found ways to meet real needs in effective, often new ways, managing conflict in the process. Pure mimicry may not work as well as it appears to.

There is also some academic research on the management of perceptions of scarcity (Cialdini, 1988; Herzenstein & Posavac, 2019; Liang et al., 2021) and (to a lesser extent) munificence (Bhatt et al., 2019) that may provide starting places for practice. Scarcity is defined using the same framework as surprise: When our expectations about what we think we need (real or imagined) don't seem to be met, we experience scarcity. The munificence of ecotones may also partly explain why they are venues for creative social organizing. Regardless, there may be important avenues for providing feedback and equably handling the conflicts and scarcities that occur in social ecotones. For example, applied scientists might direct social organizing toward saving "goods" while things are munificent—with future stakeholders in mind. More research is needed on the effects of scarcity on social speciation.

Regarding conflict management, however, many approaches have been examined in several areas of applied psychology and organizational behavior (Poitras, 2012; Samba et al., 2018; Schwenk, 1990). Perhaps the most relevant here is the scientist–practitioner literature on diversity and inclusion (Leslie et al., 2020; Tang, 2019; Triana et al., 2021), which has shown that working with groups to openly discuss and understand differences in underlying factors (e.g., social identities, personality) is likely to reduce prejudices (Leslie et al., 2020), while these same nonobvious differences can cause conflicts without such intervention (Triana et al., 2021). This research provides many options for science-based practice.

Creation of Social Ecotones

In addition to finding ways to moderate social hacking, there are sometimes opportunities for starting from scratch: creating social ecotones. Direct ecotone creation is demonstrated by the yarn bombers, where mimicry of youth gangs and knitting circles, often in safe and prosperous community art spaces, yielded a new form. Community conflicts were forestalled and scarcities addressed by forming a new social species in response to the creative perspective brought to a group that is motivated to manage social extremes. As with any social form, its long-term success is always in question, but by trying to moderate social pressures, manage conflict, and try fairly dramatic new ways of thinking and behaving, this group provides an example of a deliberate human hack—creating a social ecotone.

We need to know more about how to deliberately create ecotones. A small body of literature on self-managing team selection may hold important clues for this (Jones et al., 2000; Klimoski & Jones, 1995). Empirical examination of the processes and variables involved may follow quite clearly from what we know about blending groups with different identities, but processes may be particularly complex when "starting from scratch" (see Dokko et al., 2014; A. Lee et al., 2018).

Unlike entrepreneurial businesses, government institutions only get started once in a long while and may help demonstrate the potential delicacy of ecotone creation. In democratic societies, new government bodies usually rely on the agreement of many parties to their founding and who support their continued use. Like businesses, these organizations address real needs and have to manage intergroup conflicts just to be formed in the first place. In fact, governments are set up to deal with essential human needs that are rife with conflict and cannot be profitably managed in a market. So, for example, governments manage armed conflict with other governments, handle the distribution of essential resources (water, electric power, roads, and other utilities), and regulate the use of the lands under their control. Government institutions also take on areas of great need that cannot be met with a profit-based approach, such as providing basic resources to people who have no current resources.

Perhaps most interestingly, governments always have some means for adjudicating disputes. None of these approaches are perfect and, in fact, are often the reason governments fail. But over and again, successful governments have tried new social forms for managing disputes. For example, the desire for safety and social order often runs afoul of the behavior of people in thrall to addictions. Instead of throwing drug users into prisons, some local governments have founded "drug courts" (new social ecotones for users or offenders, citizens, and government representatives), created public health care to treat addictions, and invented other social forms to reduce the disruptive effects of addiction.

Other kinds of social species may also provide guidance for ecotone creation. Unions, cooperatives, clubs, religious organizations, hospitals, and other forms have evolved to organize people to address scarcities and support attempts to hack social-evolutionary pressures. To the extent that the social identities associated with these new forms take hold, they are the likely platforms to form ecotones; apply knowledge about mimicry, scarcity, and conflict; and find other ways to manage social speciation in the interest of a sustainable future (Bhatt et al., 2019; Sanchez-Burks et al., 2015).

New and Universal Communication

Managing conflict often relies on managing scarcity, but both rely on adequate communication. In an example of the success of communication in a social ecotone, Jared Diamond (1997) provided an analysis of pidgin languages. Pidgin languages are the result of two or more languages blending into a commonly shared language that relies on words and concepts from both original languages. Although such languages are a clear example of creative restructuring at the margins between cultures, the psychology behind pidgin languages has received limited attention (Jackendoff & Wittenberg, 2017). As with all language, core human motives are almost certainly at work here, regardless of the cultures that are blending. Also, given their inventive nature, new languages are probably not aimed only at-risk prevention. New languages blossom where scarcity is held at bay, following from the conditions of social inventiveness. This may make language a key avenue for managing conflict in a social ecotone.

But the evolution of a new language in the ecotone is only one of the several possibilities of language invention at the social margin. In some instances, one language "conquers" another. The yarn bombers have used a different strategy. Their work relies on a communication medium that is neither spoken nor invented but can be understood by broad groups of people across cultures: art. Painting, music, theater, and all the arts can serve as "surprising" communication across more munificent social ecotones. Yarn bombing, nationalization of lands in the American West, and the activities at the Rendezvous all used the arts as a motivator and communication channel.

As frivolous as this may sound, parties are one of the great blenders of people. One of the most successful city council members in Springfield, Missouri, argued that a good party is a solution to many problems. It is no coincidence that music is the language heard by people who attend parties. To the extent that their social identity is realized in this music, it is a place where people experience a comfort recalled from other "safe," munificent group experiences. And voila! People start to speak to each other as if they were part of the same species.

Other times, completely new languages are deliberately invented solely for the sake of social invention. These are not the same as spoken languages and include everything from ancient cuneiform to modern mathematics and computer codes. These invented languages do not happen at the social ecotone but are developed for the social inventions that arise in the ecotone. Only a few members of a group need to develop such languages—a factor that bears remembering when we consider how to apply the language of psychology to managing social speciation. Scientist–practitioners often find value in learning and using the jargon of the groups with which they work.

SUMMARY ANALYSIS: HACKING THE SOCIAL SPECIES HACK

Again, attempts to reorganize socially arise from the same key factor that characterizes our physical hacks: moderating the extremes in our environment. Social hacks aim to moderate behaviors in the social environment.

Just as temperature controls in our dwellings have moderated the effects of extreme heat and cold in the environment, hacking our social speciation allows us to manage social interactions in new ways. The most extreme forces of social environments are scarcity, conflict (which often arises from scarcities), and the inability to communicate. Most social groups deal in one way or another with real or perceived scarcities and the conflicts that arise from these scarcities.

If we want progress on issues such as sustainability, we need to support social inventions that manage scarcity, provide some common communication and social support, and encourage surprise and peaceful conflict among different social species. As yarn bombers demonstrate, we may not need to invent entirely new social forms but rather experiment with combinations of existing forms. If social ecotones are where such experimenting occurs, finding ways to introduce different social species and create ecotones is a logical way to ignite this process.

To be clear: Conflict should be assumed in social ecotones. Managing surprise, which leads to conflict sometimes, may suffice. But preparing for conflict makes sense, as does having means in hand to manage it. For some, conflict can be turned inward to elicit developmental change. This may explain the success of efforts based on prominent minority leaders' argument for open, civil discussions about cultural differences (Hughes et al., 2020). For others, conflict can be used as an aversive contingency—an outcome so undesirable that people alter their behaviors to avoid it. Some flexible adversarial forums, such as well-managed judicial systems, mediation, and restorative justice methods, are already available.

We may also need to create larger scale, hybrid social structures to manage conflict. This is where special languages like math and computer code allow for managing speciation processes by following the lead of "experts" who speak these languages. Imagine the tech revolution if everyone had to learn every kind of coding to be able to use basic functions. Instead, we have allowed ourselves to be led by people who have specialized knowledge of these languages. Similarly, the peer review process in science is an example of a social species that relies on a few "experts" for its successful advancement of scientific knowledge and with some success at reducing conflict.

But reliance on small groups of experts has its own problems of risk, conflict, and communication. And this brings us back to the difficult question of how to extend innovations in social ecotones where some people are unable to take a developmental perspective. In the next chapter, I argue for several approaches to this, all reliant on some understanding of developmental differences, managing conflict, and stakeholder engagement. Where some people in an ecotone have not reached the developmental perspective required for change, psychology provides various kinds of social influence that will help manage the conflicts that arise.

7

Using Social Speciation on Purpose

With these variables for managing social speciation in mind, we can try to answer the big question: How do we deliberately manage social speciation for a sustainable future? Let's assume that social ecotones are the best platform that we currently have for efficient, if not instant, change. We know that putting people with different social identities together in a fairly safe, munificent place to face a shared challenge makes it likely that the conflicts they encounter can be effectively managed, and the solutions they devise will last, so long as these rely on common human motives. Though developmental change is more likely in these circumstances and may help in developing new social identities, shared mental models are the defining characteristics of new social species. We have defined the challenges of sustainability in this way in the previous chapters; it's now time to use applied psychology to help direct and regulate speciation processes toward sustainable choices.

---

https://doi.org/10.1037/0000296-007
Sustainable Solutions: The Climate Crisis and the Psychology of Social Action, by R. G. Jones
Copyright © 2022 by the American Psychological Association. All rights reserved.

Trying to manage all the moving parts in the social speciation process looks daunting, even to experienced organizational scientists. It would be nice just to have a big party that blends all kinds of people together. People from around the world would listen to music that relates to their experiences, with everyone feeling safe and creative, and the rules of conversation would be set up and followed to avoid destructive conflict. Plenty of good things would be on hand to attract and maintain their involvement. Individual developmental (i.e., "transformational") change will be an explicit, well-supported purpose of the gathering.

Of course, this approach by itself is not likely to manage eco-hacking, but it has been tried with some success. Burning Man is a recent example of a mixer where thousands of artists, technologists, and even some new social species gather for a 10-day campout on a dry lake bed in Nevada. The organizers aim to incite "transformational change" (Beaulieu-Prévost et al., 2019) through celebrating new and unusual creative expression (K. K. Chen, 2012). Burning Man's organizers prohibited commercial activities, effectively separating people from the social environment most of them experience during their daily lives, but they deliberately encouraged all kinds of "surprise" at this huge, 10-day party. The festival has spawned new ideas and approaches to life, some of which have made it to the mainstream of art, fashion, advertising, engineering, and science ("No Spectators," 2018), if not social organization (Austin & Fitzgerald, 2018; Hoover, 2008). Not surprisingly, formal conflict management processes have also found their way into the event (Hedeen & Kelly, 2009).

The Harlem Cultural Festival, Woodstock, Lollapalooza, and other examples point to what is missing from just creating a social ecotone (like Burning Man) to manage the creation of new social species. Burning Man, in particular, is peopled by a fairly exclusive group of mostly White, highly educated, mainly Bay Area residents. Although there is diversity in the kinds of creative work these folks do, Burning Man is not so much an ecotone as a gathering of like-minded people away from their "typical" niches. While there are plenty of people who are open to experience, socially supported, and relaxed at Burning Man, there are still some important conditions that need to be in place to create lasting social speciation.

Returning to the flight analogy, well-fashioned social ecotones like Burning Man provide the form for social speciation but rely on individual motives for the other ingredients that can spur attempts to "fly"—to form new social species. Getting away from commercialism to participate in high-impact, collaborative creative activities is certainly enough to spur some individual developmental change (Beaulieu-Prévost et al., 2019), particularly given the ambiguous social identity of life on "The Playa" (the dry lake bed where Burning Man is located). Even the ingredient of self-conscious mimicry may be at work here. But beyond such social identity conflict, social speciation relies on a problem or opportunity shared by more than one group. So, presented with a persistent but manageable conflict between species and, especially, some shared need to attend to an existing problem, perhaps a more permanent Burning Man would be a good place to expect some serious speciation. The next question is: What role would applied organizational scientists play in steering social speciation toward a more sustainable future?

HOW TO SUPPORT SUSTAINABLE ADAPTATION

For scientist–practitioners, the first answer is actually an ethical caution. Until we integrate a body of research evidence that directly addresses the conditions of social speciation, applied social scientists are basing interventions on either broader principles or research that addresses processes that appear to be related. There are examples of these principled and apparently related lines of research throughout the previous chapters. Few apply directly to social speciation, and I review a few more in this chapter. The point here is that we are now venturing into methods for which, unfortunately, there is almost no applied research base. Instead, the research base directly examining "social speciation management for sustainability" will rely on the future work of scientist–practitioners.

Making things even riskier, these pioneer scientist–practitioners will be venturing into methods that may make it possible to control consequential social-evolutionary hacking. There are many psychological approaches that can be tried in support of this aspiration, but most that have proved effective in similar circumstances rely on one group of actors deliberately trying to control others' behavior. These carry high stakes for future humans and increase the potential for conflicts, some of which will be hard to foresee in a new area of practice. As important as it may be to steer people away from the use of fossil fuels, for example, the importance of this end does not necessarily justify the use of every psychological means available. Fighting for government funding of commuter rail lines using psychology-powered media campaigns sounds innocuous but may cost poor people their neighborhoods once such commuter lines are successfully developed.

Taking such conflicts into account early in consideration of this course of action is both practical and ethical. Weighing the potential damage that can be done by the general strategy—deliberately trying to manage speciation in social ecotones—against good things that it might achieve is made difficult by the lack of previous research directly relating to this endeavor. This absence makes it difficult to predict the consequences of interventions. Relying on principle alone doesn't work here either. Given the complexity of some of the issues involved, prediction will be problematic in any case, and new ethical questions will arise (Bhatt et al., 2019). Just because scientist–practitioners will be providing a voice for an unheard stakeholder group (future people) does not mean that this group's interests always overrule other stakeholders' interests.

The good news is that, sometimes, doing the right thing and doing the most effective thing are one and the same. Commuter rail lines along existing rail and road corridors that are free for low-income residents, for example, are often feasible, ethical, environmentally responsible, and effective in practice. But when a course of action is likely to cause some harm—which is the more common circumstance with psychology—it is a fundamental responsibility to exercise ethical caution (Jones, 2020, p. 301; Lowman, 2013). Here are some questions to think through before taking a course of action:

- Are you trying to change people's thinking or just their behaviors?
- Do you have the informed consent of a representative portion of the people whose thinking and behavior you are trying to change?

- If you cannot get consent, would the kinds of changes you are anticipating likely cause any harm to this or other groups? If so, what are the specific kinds of harm that you might cause, and how can you reduce the chances of these harms?
- If thinking is your target (i.e., developmental change or new shared mental models), do you need to change everyone in a group or some portion of the group? For example, if you are trying to change the thinking of a conventional or authoritarian religious group, you may need to change the thinking only of their leadership to create behavioral change in the larger group. If you are relying on this small portion of the group, what are the potential negative consequences for all group members?
- Because social speciation and developmental change rely on conflict, what measures can be put in place to manage this conflict? What is being done to reduce the likelihood of destructive behaviors (including hate speech and other destructive language)?

There are many more questions, but these are a good starting place. As for the more general principle-based strategy for dealing with potential harm, it is clear that conflict management is essential for devising new social species. We're dealing with social ecotones, after all, where people with different social identities meet. These are both venues for creative social speciation and profound, intractable conflict (Diamond, 1997). If managed expertly, conflict may lead to profound developmental change (Dokko et al., 2014; Sanchez-Burks et al., 2015); managed poorly, such conflict leads to violence (Leidner et al., 2013). Therefore, scientist–practitioner competencies should certainly include conflict resolution knowledge and abilities (see DeChurch et al., 2013).

A Second Caution: Competence

Briefly, I am an industrial and organizational (I/O) psychologist. Although this gives me some credibility as an applied methodologist, my knowledge in some areas of psychology is only slightly better than rudimentary. Along with the ethical cautions posed so far, expertise in various areas of practice other than I/O psychology is also essential to our success.

In fact, such competency is also an ethical imperative (Lowman, 2013). I hope for your help. Many of the readers of this volume are themselves experts in some area of psychology or, as students, are developing expertise in an area. The margin between expert and knowledgeable amateur is a fertile place for new ideas—some of which are likely to break new ground and lead to new inventions. I know that my perspectives as I started the journey toward expertise were sometimes silly—but also sometimes the seeds of ideas that turned out to be enduring. So please watch for ways to organize your social actions in ways that deal effectively—and ethically— with the variables needed for social speciation. My grandchildren need your help. And although I have the ingredients ready (see Exhibit 7.1), the recipe requires further consideration by other people with expertise— like you.

Applying Psychology to Ecotone Management and Social Speciation

According to the rapid increase in individual comfort and safety over the past 100 years or so (BBC, 2010), social invention has been booming. People at the margins of social contexts (minorities, prophets, people with "mixed" social identities, various expatriate communities) appear to have inspired widespread, even developmental change (cf. Drescher, 2003; Tang, 2019), along with the creation of new social identities and processes (cf. Krupat et al., 2013).

To harness this boom and turn it into a more sustainable future, it helps to look at a recent, successful example of social speciation. Virtual teams (and teams more generally) are now a broadly used, growing example that has spawned a substantial enough research base to draw some conclusions. Research on virtual teams (Brown et al., 2021; Lines et al., 2021) offers important insights into the development and functioning of this type of social form (Lines et al., 2021), which is based in a delimited social ecotone. *Virtual teams* are standing groups charged with trying to manage some problem in their host organization without the benefit of direct,

Exhibit 7.1

Ingredients for Novel Social Speciation

Situational Factors

- A fairly constant (long-lasting) social ecotone
- Relative safety
- Some scarcity or conflict that is not life threatening in the short term
- Some expectation of success

Actor Factors

- Curiosity: high openness to experience
- Relaxed and open to feedback and change
- Willing to embrace complexity
- Has some situational expertise
- Perceived resourcefulness
- Readiness: nonauthoritarian, has long-term perspective, has later stage social development

Group Process Factors

- Social support for learning and change
- A conflict or scarcity shared with other social species in the ecotone
- Conflict management processes in place
- Language and emotive language tools
- Perceived shared goals
- Facilitator, mediator roles
- Focus on behavior

face-to-face contact. They typically require extended interactions among people with diverse knowledge, skills, and professional backgrounds using technology.

Given the problems of managing conflict, shared motives, and mental model development in ecotones, it is not surprising that this new social species suffers from problems in all these areas (Breuer et al., 2016; Brown et al., 2021; Lines et al., 2021; Rosen et al., 2007). More helpful is that efforts to develop leadership and shared mental models have enhanced the success of these and other types of teams (Breuer et al., 2016; Brown et al., 2021; Lines et al., 2021). This research holds considerable promise for managing social speciation, more generally, through several means.

Applied psychologists already provide support functions that may prove valuable for managing rapid speciation in other social ecotones. Psychologists working in applied areas have developed successful systems to manage many types of social structures and, as with virtual teams, may be able to direct efforts toward sustainable speciation. I/O psychologists, in particular, engage with decision makers to support decisions about how to reform everything from small working groups to large multinational organizations. Like I/O, educational psychology applications are used to support basic decisions about selection and training. In terms of social ecotones, these assessment methods support decisions about who to include in—and exclude from—work and educational organizations. These methods are used widely to help select students, hire new employees, manage developmental boundaries, and decide who to educate in what areas.

Note, though, that these methods control the boundaries and behaviors of an existing social group rather than constructing a social ecotone, then developing new social species in that ecotone. Partly as a reaction to the discontent these traditional selection methods have bred, in recent years many schools have designed selection methods that deliberately include social differences (Bowman, 2013; Pike et al., 2007), while others have developed training that supports this deliberate ecotone creation (Bezrukova et al., 2016). In the absence of future stakeholders available for hire, however, selection systems are unlikely avenues for direct sustainability interventions.

But these efforts do point to valuable frontiers for practice. For example, diversity and inclusion interventions (discussed in the previous chapter) are sometimes developed by applied psychologists (Leslie et al., 2020; Triana et al., 2021) to manage conflict and integrate social identities. These often include a role-playing process that could be adapted to include a "future stakeholder" role. Likewise, team development interventions (Lines et al., 2021; Tang, 2019), boundary spanning leadership development (Jones, 2020, p. 280), and managing organizational strategy (Pulakos et al., 2019) all have value for forming shared mental models and, potentially, for enhancing sustainable social speciation. Assessment will remain an important, flexible tool, perhaps expanded for making decisions about the creation and management of ecotones.

Targeting for Social Speciation

Continuing with the team development example, the research evaluating attempts to create self-managing teams are instructive—at least when later stage individual development of members is assumed. Self-managing teams became a popular approach to structuring work in the 1980s, and they have had enough success to still be found in many workplaces today (Mesmer-Magnus et al., 2017). These are typically multidisciplinary groups that devise their own mental models of the problems they have been constituted to address, then arrange their activities according to these models. In other words, these groups are responsible for inventing themselves—forming their own social species around a shared mental model.

Early evaluation research uncovered issues that affect the transition to team structures (Jones & Lindley, 1998; Kirkman et al., 2000). These discoveries may also be important to managing social speciation more generally. Findings from the first group of studies will not come as a surprise here. After extensive data gathering and analysis, most of the concerns of new team members across several large organizations related directly to social support. There is considerable uncertainty that accompanies trying to address a new problem in a new way, and having the support of managers and coworkers during this process affected the teams' ultimate success (Jones & Lindley, 1998; Kirkman et al., 2000).

Similar research on methods for assessing and hiring team members uncovered several psychological differences that support—or impede—self-managing groups. First, this research points to the importance of being able to deal with people developmentally different from oneself (Bell, 2007; Jones et al., 2000). On average, groups performed better when their members shared certain personality traits (often defined in terms of developmental differences) and when they were consistent among themselves on factors related to social interaction and authoritarianism (Jones et al., 2000). Many of these individual differences relate to differences in moral and social development, in particular (Jones, 2020, p. 173), both of which have relevance to a sustainable worldview. Second, groups that can recognize and rely on differences in each other's expertise perform better across circumstances (Grutterink et al., 2013; Zemel et al., 2011). This may partly be the result of drawing on different sources of information about their work environment, but it may also be an indicator that most or all members of the group are able to accept feedback and learn from one another. Whatever the reason, group members who don't see value in others' viewpoints are likely to be problematic for establishing shared models.

The conclusion from this is that different approaches to social interaction can impede groups' attempts to redefine themselves. This points to an essential factor for managing our hack. Successful social speciation needs to account for such developmental, expertise-based, and feedback management differences in the parties to the speciation process.

CONFLICT

Where do these differences between group members have their impact? The answer is still unclear, but a developmental perspective may help—and it points to conflict as the culprit. Recall that authoritarianism is indicative of an early developmental stage. Members who are higher in authoritarianism are looking for an authority to answer their questions—which is not where you are likely to find meaningful answers when you're in the early stages of a self-constructing group. Those who are low on authoritarianism (have advanced into later stages) are required to "think like their younger selves" to relate to teammates who are still in earlier stages.

So, trying to instigate innovation relies on developmentally advanced group members who are not only willing to deal with uncertainty but are also willing to deal with the reactions to the uncertainty of group members who are less well-equipped to deal with it.

A common educational practice helps to demonstrate this problem. Anyone who has, for any period, watched children of different ages try to interact knows that there will be conflict. This should come as no great surprise because different developmental stages rely on entirely different understandings of how the world works. People who provide services to children have been separating them into age groups for a long time. In fact, an early assessment of developmental differences was devised by Alfred Binet for just this purpose—to measure "mental age" so that the French department of education could make class placement decisions (Hothersall, 1983). Binet based his method on the common errors of thinking that children make at different ages—now called stages of cognitive development (Smith, 1994). Depending on the questions children answered correctly on Binet's test of mental age, they were assigned to different grade levels. Placing people with different developmental understandings into different groups is sensible when trying to educate children.

But it makes things considerably more complicated to deal with adult developmental differences. Putting two people together in the same working group who differ in their fundamental assumptions about the source of correct information (e.g., uncertain science vs. "certain" authority), their basic perceptions of time (e.g., everlasting life vs. abrupt mortality), and bases of human motivation (e.g., "fixed" in people's character vs. changeable based on situations) is a recipe for conflict. Occasionally, such conflicts may be managed well enough to cause individual developmental change. Otherwise, those of us who would like to manage social speciation need to make allowances for this set of ecotone problems.

Managing Conflict With Perspective Taking

A good place to start here is with an understanding that, although later stage development in every person in an ecotone might be a good aspiration, trying to reach this aspiration is time-consuming, painful, and extremely unlikely, especially given the idiosyncrasy of this kind of change (see Chapter 5).

Instead, it makes sense to follow the I/O psychology assumption that conflict is an inevitable part of any organizational change. Books are written on the topic because managing conflict is essential to the success of scientist–practitioner efforts. Even this does not mean we expect to do away with conflict. It means trying to direct conflict toward constructive ends or at least deflecting it away from essential processes.

Constructive conflict management starts with a careful analysis of what conflict is. During his highly productive career, Dean Tjosvold shed light on the nature, causes, and management of conflict. He defined it in terms of people's perceptions about whether their goals are shared (Tjosvold, 2008). Cooperative processes start with assumptions group members have about one another's motives. If you can establish that the parties to a conflict share some motives, there is an opportunity for cooperation. This should sound familiar because we have already seen that shared ends help develop social species and that differences in developmental stage are defined in terms of different assumptions about motivations.

Tjosvold's (2008) idea of constructive controversy provides a cooperative method for managing conflict, if not necessarily instigating developmental change. Identifying shared purposes is the core concept of his method. Certainly, when people meet in a social ecotone, we can assume that conflict will arise. But tempering this conflict into cooperating to reach some shared end makes a lot of sense if it can be accomplished. Getting people in whatever developmental stage to try to cooperate relies on the assumption that there are some shared motives at work, even if people in the group don't share fundamental assumptions about human motivation. Creating a perspective that includes some shared purpose that can only be achieved cooperatively is an essential ingredient for social speciation.

Emotive Aspects of Conflict

The first method for managing conflict comes from clinical and counseling psychology and targets emotional communication. A defining characteristic of conflict is that it involves both motivation and "emotivation."

Whether people are aware of this or not, our limbic system (also known as the old mammalian cortex) gets involved in our efforts to accomplish some outcomes and avoid others (Elliot et al., 2013). In his controversial article, "On the Primacy of Affect," Robert Zajonc (1984) argued convincingly that we are, like all other mammals, emotive animals first and rational beings only after this. Kahneman's (2011) research also demonstrated quite clearly that we are not "rational." Certainly, some people deny their emotive responses because these are often seen as childish, antisocial, and generally annoying. But the evidence is quite clear that we all have emotive episodes (Weiss & Cropanzano, 1996)—usually short bursts of brain activity in the limbic system in response to events in the world around us that have relevance to things we value (Jones, 2015).

Individuals develop different ways of coping with these emotive episodes. Most of the youngest members of our species express their emotive states spontaneously. Others of us, trying to fight or ignore the "sinful" motives underlying these emotive responses, do our best to deny the existence of the emotive response itself. Here, we are usually concerned with behaving in the way our current social identity (e.g., culture, work group, customers) requires. Some portion of the adult population has also moved to a stage where they are able to take perspective on what I call their "inner doggie"—confidently managing the appropriate behaviors while still allowing their limbic system (inner doggie) to do what it does. A figurative leash can work well.

Conflict resolution research suggests that this "allowance for emotion"—but not necessarily the behavior following from it—is an essential first step in conflict management (Jiang et al., 2013). Counseling psychologists and other professionals who deal with conflict have a truism: Conflict resolution starts with verbal recognition of the legitimacy of the emotions being expressed. "I can see why you feel that way. I would feel the same way in your situation" is a common first response to emotion in conflict. This "patting" of your inner doggie is thought to reduce focus on external triggers (especially relationship-based) and conflict and focus instead on solving the problem at hand (de Wit et al., 2012).

Another component of this emotive management is the social support function that it can establish. We don't just experience emotions—we also express them, often without realizing that we are expressing them. As with other social animals, these expressions communicate motives to others (Wharton & Erickson, 1993). So, when we meet someone on the street, a smile is usually interpreted as "happy to see you" and a frown as "not so happy to see you." Making people aware of the effects of their emotive displays on others is one of the important approaches to establishing "us and them" differences—inclusion and prejudice are signaled by emotive displays. If we are trying to establish cooperation and manage conflict, smiles can go a long way.

Recognition of Expertise and Conflict Management

After this, perhaps the most constructive end scientist–practitioners can hope for in conflict management is the advancement of learning or development. This is the second approach supported by current practice. We have already seen that conflict between an individual's social identities is the crucible for development—when managed well, it is also a spur for social speciation. Such development and speciation are not likely in the "hot" moment, however. Even if the parties to conflict eventually recognize another party's viewpoint and integrate it into their cognitive maps, this is more likely to happen when they are "relaxed and supported," which is not what is usually happening in the emotional heat of interpersonal conflict.

But helping people to understand each other's actual (vs. assumed) motives is easier when emotions are acknowledged and understood as signals for motives, especially when different developmental stages are underlying the conflict. In fact, aside from the knowledge of group members' motivations as a means for conflict management, recognition of member expertise is a potentially powerful avenue for directing speciation toward sustainability. Establishing members' various areas of expertise early in group interactions has shown promise as a way to enhance group effectiveness and cooperation in various settings (Bazarova & Yuan, 2013; Grutterink et al., 2013; Joshi, 2014; Zemel et al., 2011), perhaps partly because it clarifies or even creates shared motives. Given one or more members who either bring expertise regarding questions of sustainability or are educated about them during group formation is a natural way to interject dialogues about sustainability into groups' identities and shared schemata. This is a potentially fruitful avenue for further research.

Back to conflict: Developmental change in social identity may be unusual mainly because it involves these aversive emotional interactions. Not only is change sparked by conflict—which often includes aversive emotions—but so are withdrawal, aggression, and violence. Tjosvold and many other organizational scientists are skillful in the use of emotive language, conflict management tools, and approaches to difficult choices so that they can manage just these sorts of aversive, conflict-laden social interactions. These keep conflict in the discussion but keep it from becoming so aversive that it overwhelms the conversation. The perspective that certain uses of humor can bring to these difficult interactions can also help (Metzger, 2014; Robert & Wall, 2019), though there are certainly other skills (e.g., empathic listening) that are needed to support humor as an approach to interpersonal conflict management (Jiang et al., 2013). Successful, sustainability-oriented speciation probably relies on the skillful balancing of conflict and creativity—keeping conflict regulated and directed toward sustainability but not letting it become so aversive that it squelches creativity (Sanchez-Burks et al., 2015; Tang, 2019).

Behavioral Aspirations

Given that this kind of skills learning is a common means to manage conflict, it's fair to suggest that cognitive change is more realistic and easier to attain than development. Also, applied psychologists who are trying to manage an ecotone are responsible for managing conflict anyway. So, the next step in conflict management is to consider just what other ends—short of developmental and cognitive learning—we might hope and plan for accordingly. For example, having put teenagers with different social identities (two high school football teams) together at a Friday night game, school officials and parents might aim to hold extreme behaviors in check until emotive episodes have subsided. Framing desirable outcomes in this way—as behaviors—is both a realistic goal and a good general rule.

It is also essential to the third tool for conflict management. This is to focus on behaviors. Discussing behaviors rather than causal explanations for these behaviors (i.e., attributions) tends to avoid conflict and enhance learning (Smither et al., 2005). Far too often, people rely on folk theories of psychology to explain a behavior. For example, I might see someone cheering for a political candidate who I know has demonstrated in many behavioral instances that they are not worthy of the public trust. I might assume that the person cheering for this candidate is stupid or any number of other pejorative attributions that can be used to explain their "cheering behavior." However, focusing on the cheering behavior allows for a host of other possible attributions—not just the "stupid" attribution. For example, if this is happening next to the person's spouse or parent, I might see that cheering is an important behavior for maintaining good relations with this beloved other person, regardless of the actual beliefs of the cheering person.

More important, this focus on behavior applies to how we interact with other parties to a conflict. Following this example, I might be standing next to this cheering person, who is my friend. If I fall prey to the attribution, I could do permanent damage to this relationship by openly calling my friend "stupid" for having cheered for a reprobate. Instead, I might take the person aside and mention the behavior I observed: "I saw you cheering for ___," then leave them to respond.

It's remarkable how successful this approach has been—so much so that many studies have shown how such "behavioral feedback" leads to successful outcomes—learning, improved performance, lower conflict, and behavior change—even in high-stakes circumstances (Balzer et al., 1992; Kluger & DeNisi, 1996; Smither et al., 2005). So, stop and observe and talk about the behavior—not about explanations for the behavior.

All of this is to say that if behaviors are all that we can manage when there is conflict, we may be doing better than we think. But it also uncovers an unpleasant question for social speciation. If groups are unable to relate at the same developmental stage, is it ever possible to form shared mental models? Even with highly skilled psychologists facilitating the speciation process, it may be that some shared models will always rely on deliberately choosing group members who are at least inclined to try to be able to understand the group's common motives, conflicts, and adaptive hack. Starting with the assumption that some members of a speciating group will be unable to fully develop or understand the shared mental model the group relies on may be a fatal problem for speciation—not just because of conflict but because everyone may need to be "on the same page" for certain kinds of tasks.

Boundary Spanning Leadership

Let's assume, then, that we cannot instigate developmental change in most of the people in our ecotone. Then what?

The good news here is that people who have advanced to later stages of development are more likely to recognize when others have also advanced to this stage—or not (Gabenesch & Hunt, 1971; Jones, 2015). They may have another advantage, as well: being able to relate to the people at earlier stages because they have been in these earlier stages themselves. If you have reached a later stage and are willing to take on the task, you are in the unique position to help others develop some understanding of a complex mental model and help thereby to direct speciation toward sustainability.

This does put a burden on group members who are further along developmentally. For example, their low reliance on authority, long-term temporal perspective, and understanding that people are changeable (all developmental differences) give them an advantage but also saddles them with the difficult task of trying to instigate change in others. On the plus side, this could also mean that to manage social speciation effectively, sometimes only one member of a group needs to have advanced to a later developmental stage. Providing members who share this perspective with "boundary-spanning" skills is one way to prepare them for taking on this role—trying to manage differences between two or more groups with which they are engaged (Dokko et al., 2014).

Some of the core methods for conflict management will help with this role. Leaders in later stages can focus on behavior (Kluger & DeNisi, 1996) and manage conflict by using socioemotional support and shared tasks (Burke et al., 2006; DeChurch et al., 2013) to try to advance others' cognitive understanding—if not development. Guiding people through the conflicts that result from novel experiences they encounter in social ecotones is one of the main avenues toward the shared mental models on which social species rely. It is also a way to frame the group process commonly referred to as leadership (Guenter et al., 2017; Wilke & Thornton, 2019).

But rather than saddling this one group member with advancing others to later stages, there are other approaches to directing a group toward a shared identity and mental model that includes a sustainable future. This is where other forms of leadership and social influence can come into play. Psychological science provides a growing list of successful social influence tactics to help at least change the behaviors (if not the developmental stages) of others (Burke et al., 2006; Hornsey & Fielding, 2017; Vroom & Jago, 2007). Based on an understanding of developmental differences, some perspectives on leadership and social influence likely affect social speciation.

The Nonbelievers

It will have occurred to some readers that, because of risk (and conflict) aversion, a substantial portion of the population will resist developmental change and social speciation—often violently. This resistance barrier (Kirkman et al., 2000) goes beyond mere conflict, so direct evidence regarding effective ways to overcome resistance to social speciation is needed. A large portion of the human population carries a strong potential for resistance to the idea of managing social speciation to account for future stakeholder representation. It makes sense to start with an ethical analysis because there is little evidence to guide applied scientists' efforts.

Here, my experiences as an elected member of a nonpartisan city council provide an unusual perspective. On the one hand, decision makers could simply let the "resisters" live with their choices, effectively leaving them out of the speciation process. In decisions where the majority rules, this is sometimes the initial effect. The social world has changed, becoming far less homogeneous and "comfortable" for those who rely on their old, often culturally embedded heuristics. As with any adaptive change, those who fail to adapt successfully fail to thrive. They tend to live desperate lives, die sooner, and fare poorly compared with those who are in privileged positions to begin with. Here, privilege is defined as having resources of whatever sort are necessary to adapt successfully.

On the other hand, decision makers could expend valuable time and resources to lessen the impacts of change, trying to adapt the speciation process to include the interests of people inclined to resist such change. On the basis of a utilitarian ethic, political entities have tried to make a place for people who fall behind. The theory underlying this approach is that by providing social goods to people in less privileged situations—free education, assistance with food and housing, and so on—society will at least minimize the cost of adaptation. Giving less-prepared people "part of the goods" is thought to make them better able to adapt and less inclined to resist change in the longer term.

Unfortunately, in both scenarios, there is likely resistance from both the privileged (some of whom want to keep their privileges intact and others who value helping others more heavily) and the less privileged (who tend to resist change of any kind due to lack of resources to manage uncertainty). The third approach deals with both kinds of resistance. Where the first two approaches rely loosely on evolutionary biology ("survival of the fittest") and political theory (utilitarian ethics), this third approach has the substantial advantage of relying on human psychology. But like most solutions that take complex realities into account, it is complex and, therefore, difficult to implement.

Following the government analogy further, this applied psychological approach treats resistance as a sort of "road-building" problem. First, it is essential to draw a map of the human "territory" we hope to traverse. Fortunately, Chapters 4 and 5 provide the framework for this, and it starts with the flipside of "readiness" factors (curiosity, imagination, and social support). This is the psychology of desperation—of people who feel unprepared to deal with changes. Social identity and developmental differences provide some guidance for efforts to win the desperate. Depending on the developmental phase, the social–psychological strategies of authoritarian, conforming, and minority influence attempts may prove invaluable for implementing shared mental models across disparate social groups—despite conflict and the resistance it breeds.

Authoritarian Influence

As children, all of us relied entirely on some "end-all" authority when we were, in fact, entirely dependent on caregivers. Some are unaware of any other social arrangement, often because they are still doing it. For most adults, though, the simplistic assumption that "everything will work out in the end" if we follow "the rules" set out by a powerful other eventually gives way to reality. We become aware that the simple either–or contingencies that underlie the rules don't work, that certainty is an illusion (except for death and taxes), and that the rules are not the same for everyone, depending on where they grew up, their social identity, and a host of other factors.

Eventually, most become aware of what famous social psychologist Robert Cialdini (1988) called the "authority = right" heuristic. Most come to understand the inadequacy of reliance on authority to navigate the complex problems in a complex world. Those who manage to develop past this stage also come to recognize those moments when our fellows fall back on authority—on some god document, a living religious or political figure, a parent, or some other "ultimate" set of answers to life's great uncertainties. We come to understand that the world's complexities can seldom be reduced to a religious tenet, command, or other pronouncement from any authority. But we also come to understand that others may need this kind of authority for their peace of mind.

A minority come to realize just how uncertain the future is (Jones et al., 2016). This same unpredictability and complexity are frightening enough to another minority that they revert to the old authorities, despite being aware of the inadequacy of an authoritarian approach to life's challenges (Oesterreich, 2005). These folks rely on the authority = right heuristic, even though they know that the safety they seek is illusory. Not surprisingly, measures of adult authoritarian worldview are related positively to measures of anxiety (Jost et al., 2003; Singer & Feshbach, 1959) and negatively to scores on mental abilities measures (Heaven et al., 2011). If the world is frightening, and I don't have great cognitive mapmaking skills, I am more likely to fall back on this authority = right heuristic—likely to my detriment (Heaven et al., 2011).

Given what we know so far, anxious people are also less likely to try to devise evolutionary hacks by themselves. Inventing heating systems, food processing, and other means for moderating environmental pressures is more likely with people who are not as anxious and have resources (time, knowledge, methods) to analyze some complex problem.

But authoritarian thinking poses an even more important impediment to social invention. Social invention occurs when different social groups interact peaceably. Authoritarianism is associated strongly with social prejudices—seeing people with visibly "different" social identities as less deserving, less moral, and less able (Cozzarelli et al., 2001; Heaven et al., 2011; Peterson & Zurbriggen, 2010). Conflict may be quite a bit harder to manage and coordinating between groups less likely to begin with when there are authoritarians in the mix (Jones et al., 2000). Unless their favorite form of social support is somehow "at hand"—in the form of an apparently confident, certain authority—people with authoritarian worldviews are not likely to budge. In fact, passive and occasionally overt hostility are likely results of attempts to change things they see as relevant to their fears (Oesterreich, 2005).

Having the person "in charge" be developmentally advanced may not help with this. From the authoritarian worldview, such leadership can appear uncertain and socially "different" and therefore likely to invite conflict (Cozzarelli et al., 2001; Peterson & Zurbriggen, 2010). Nicolo Machiavelli's *Discourses* (Walker, 1971) provides a hint for dealing with this problem. He argued that knowing how to manage appearances is an essential skill of leadership. So, where developmental differences exist between leadership and other group members, it may help for the leadership to appear quite certain but to act in accordance with the uncertain realities of a situation. In other words, the leader pretends to the social identity of authority when speaking to authoritarians to open the door to change. There is important work in the leadership literature that skirts this issue (Kessler & Cohrs, 2008; Luu & Djurkovic, 2019; Moore et al., 2019), but it may be a key component of getting desperate people to take nonviolent action in the face of environmental conflict (Fritsche et al., 2012). In simple terms, developmental change is unnecessary if you can find ways to get authoritarians to "follow the leader" toward sustainable ends.

An example of this is the reactions of authoritarians to ecological laws in Germany. A study in the 1980s, when these laws were passed, showed that authoritarians were resistant to these and voted against them in a large plurality. Ten years later, a similar survey showed that authoritarians were among the most vehement defenders of the "existing" laws (Reese, 2012). Authority was "right" even when it was contradictory.

Hopefully, most people in leadership roles will not have to resort to these tactics. Most of the readers of this book understand that authority is not always a sound basis for decision making, and the struggle to create a more sustainable future has already been taken up. In fact, it is a large minority of the population that resorts to authoritarian thinking. If applied scientists, most of whom are likely at the late stage, formal operational reasoning, can find ways to combine efforts with those susceptible to authoritarian influence, a large proportion of the population can be put to work to manage social speciation. But it seems intuitively unlikely that authoritarians will want to stray too far from their "usual" approach to things, even for the sake of a powerful authority.

Managing Conformity

It may also be possible to reach most of the population through a similar combination of late-stage and middle-stage thinkers, leaving the authoritarian minority to its own devices. This starts by preaching to the choir—enlisting people who are already at the later stage to take action. This includes many I/O psychologists' executive and managerial clients in organizations. Done well, this will reach the large minority that comprises late-stage thinkers. This leaves the largest segment of our fellows—those at the middle stage of social development—the conformers. A combination of people from these two groups and possibly inciting developmental change in earlier stage thinkers can override the desperate authoritarians.

Conforming

Adulthood is often marked by the abandonment of authoritarian epistemology (Oesterreich, 2005). Acute uncertainty about social standing, the doubtful value of authorities, and the need to find "safe" social groups lead many to look to peers to help decide what to believe and how to act. This urge to conform remains into adulthood (Cialdini, 2003). In fact, for many decisions we make during the day—what to wear, what language to use, and a host of other choices—we rely on what Cialdini (1988) described as a *conformity heuristic*. We look to "referent others"—people with social identities we emulate—to see how they act, then follow the "leader" in this way. Rather than someone telling someone else how to act, we look to successful others to make behavioral choices (Crain, 2000).

From a developmental perspective, conforming fits comfortably into the middle stage of social development, between authoritarian and more considered approaches to social behavior (Jones, 2020, pp. 188–189). Simply put, if group members are at this middle conventional stage, a successful leader can use conformity to change behavior, even without deeper developmental change. Paradoxically, leaders can use shared, existing social identities and these "typical" behaviors (norms) to get people to act in new ways, as well. These new ways of acting can eventually become new norms. This is a bit problematic for scientist–practitioners trying to effectuate permanent change because norms can be hard to assess and can be changed easily in the same way they are formed. Also, working with people in leadership roles to develop a sustainability-oriented mental map is an initial challenge. At least we can help to establish some new behaviors as "normal" for a while. Once everyone in a group starts behaving in a different way, this becomes "conventional" behavior. Meanwhile, it may be important to consider that creating a social ecotone may reduce conforming behaviors (Gaither et al., 2018)—something we yet know little about.

Normative Minority Influence

This is where minority influence can sometimes prove to be an exception. Leadership can be thought of as a minority influence. First, it needs to be understood as a group process because there are by definition at least two people involved in "leadership" and "followership." Many leadership thinkers treat leaders as the few "people at the top" of some hierarchy. Although this primate-based notion is accurate sometimes, not all leadership fits this framework. Rather than relying on this narrow definition, I frame leadership as a social minority that is trying to influence a social majority. This influence can certainly rely on authoritarian and conforming tactics, such as the approaches already discussed.

Treating leadership as a minority influence opens up a substantial body of laboratory research on minority influence as potential support for directing the speciation process (Gardikiotis, 2011; Martin et al., 2007; Wood et al., 1994). Among other important findings, this research showed that, under certain conditions, people with minority social identities can effectuate deep, enduring cognitive change on people from a majority group. Its relevance here is that *normative change* is defined by deep and enduring differences in the way people see social behaviors. Normative change can be caused by minorities who take a consistent and somewhat extreme position about a social issue (Wood et al., 1994). Under these conditions, minorities can get people to understand the contingencies of their choices in new ways.

An important example of this is the Stonewall Union, which staged protests in San Francisco over a period of years (Drescher, 2003). The public proponents for this group restated the same (at the time) somewhat controversial message repeatedly: Sexual minorities should have the same rights as other Americans. In the end, Stonewall became a famous example of people taking coordinated action to effectuate the changes in laws that followed. But this was also a classic example of a consistent and somewhat extreme minority changing people's thinking about whether LGBTQ+ behaviors are "normal" (Cialdini, 2003; Gardikiotis, 2011; Martin et al., 2007).

In its most powerful form, such normative minority influence leads to developmental change—and sometimes even to new social speciation. This brings us full circle on the problem of conflict. If you want to engender speciation, you will have to be willing to invite the conflict that comes from holding consistently to a very different view from the norm. This is the kind of courage exhibited by people such as Harvey Milk of Stonewall, Malcolm X, and other strident "outsider" voices of minorities that changed the beliefs of mainstream society. What is clear is that normative influence is conflict laden—a key ingredient in the development of both social identities and shared mental models.

It is equally clear that minority influence need not be exerted from within a social species to create cognitive change and guide social speciation. To the extent that scientist–practitioners can ethically provide support to "outsider" minority groups, the approaches described in this research may prove pivotal in directing sustainable species changes in existing mainstream social groups. Riemer and his colleagues (Riemer & Harré, 2017; Riemer et al., 2016) have demonstrated how social organizing of youth to raise awareness for sustainability can be supported by community psychologists.

As with the conditions psychologists need to consider in internal and consulting roles, applied psychologists need to gauge whether the minority-mainstream ecotone is munificent enough and whether there are adequate shared motives between the groups involved. It is also important to realize that most of the minority influence research has been done in laboratories, making it something of a blunt instrument. Great care needs to be taken with this approach, given that it is based on conflict.

It may seem obvious, too, that these strides in civil rights have both started in diverse urban regions—social ecotones where life was munificent, and most conflicts managed nonviolently. These spread to less dense places—where the ecotones were deliberately created by groups like Martin Luther King Jr.'s church followers and the Freedom Riders—but not without enormous costs.

SUMMARY

An understanding of psychology makes clear that social inventions are preceded by "deeply adaptive" thought processes and accomplished by uniquely adaptive social structures. So far, these have tended to lead us deeper into the tragedy of the commons, where individual security and comfort are gained at the expense of future survival. To use Rachel Carson's language, "managing ourselves" means taking perspective on and managing this social speciation. Gaining perspective implies developmental change, which means that managing our inclination to novel social organizing is possible using a developmental approach.

The bad news is that, although broad developmental change does happen, as we have seen in the advancement of rights for LGBTQ+ people over the past few decades, it is not the most common type of psychological change. More modest objectives are learning and behavior change. Getting people in earlier developmental stages just to act differently can be done quickly and may lead eventually to thinking differently. For this to happen, later stage group members interested in a sustainable future may need to use authority and conformity to manage norms in their groups. Such leadership is a group process that, among other things, manages the conflicting motives that occur at social margins—the ecotones where people at different developmental stages with different social identities are forced to work together toward some shared goal.

I/O psychologists work directly with many of the decision makers who are responsible for this. People in leadership roles rely on scientist–practitioners to help navigate the complex, delicate work ahead of them. This includes assessing the developmental aptitude and motives of themselves and the groups with which they work, then deciding which kind of influence might best be exerted to change behavior, alter cognitive map routes, and occasionally change people's developmental perspective. Psychologists' work often starts by spurring change in managers and executives so that they can take on this complex social role (Kossek et al., 2017; Smither et al., 2005). Providing counsel and designing systems to assess their (actual and potential) workforce is supported by helping to devise and evaluate influence tactics (usually based on survey data). To the extent that we instill a persistent motivation to consider long-term, environmental effects of decisions in both these leaders and the people they are trying to manage, we can make an important difference. Short of this, we can provide a sort of "conscience" in such decisions, based partly on a firm grounding in scientist–practitioner ethics, especially in the development of future leadership staff.

A few scientist–practitioners—usually carrying minority social identities—can also try to influence practices "from the outside." This carries enormous responsibility for trying to manage the conflict on which minority influence tactics rely.

What is certain is that social speciation relies on social influence either way—developmental change and behavior change both rely on it, as well. Understanding the psychological ingredients of social speciation (see Exhibit 7.1) will help guide this process. This starts with understanding the social context of the applications of psychology that we have visited and the inevitable conflict that accompanies change. Skillful management of social and psychological boundaries can lead to speciation that enhances chances for a healthy, sustainable future.

8

Defining Successful Social Speciation

New social species may be commonplace these days, but they probably have a high mortality rate. If we take the number of new businesses started in the United States each year as a fair indication of social speciation more generally, the failure rate for new enterprises is somewhere in the 40% to 50% range within the first 5 years (U.S. Bureau of Labor Statistics, 2020). That is, slightly over half of all new businesses survive for more than 5 years. The reasons for this have been studied and debated by generations of organizational researchers, with various explanations offered.

One explanation is that new ventures of all sorts must establish their legitimacy—with investors, customers, voters, donors, the public, and others (Amankwah-Amoah et al., 2019; G. Fisher et al., 2016; Petkova et al., 2013; Rao et al., 2008). This legitimacy problem can be framed using an applied psychological question we have already asked: Does the new social organization meet existing human needs?

To answer this question, scientist-practitioners can rely on the basic research visited in Chapter 2: People are strongly motivated to avoid risk and less strongly motivated to gain advantages (Kahneman, 2011). From this, I have drawn ideas about how to manage risk perceptions using social support and a few approaches to conflict management, with the idea of managing creative change. Social speciation—the construction of social organizations to moderate social-environmental demands—relies on understanding the risks that stakeholders are trying to avoid, as well as the opportunities being sought by their involvement. The analysis that follows can be used for identifying and managing some of the complex motivations of stakeholders in the speciation process.

An example helps to illustrate the nature of this multiple stakeholder criterion problem. If enough people feel that "bad hair" is a problem, a new hair care product might emerge and succeed. So, in the usual pattern of social speciation, a chemist, a successful hairdresser, and a local banker decide to start a new business, based on the chemist's excellent new organic formula for "curing" her beloved spouse's bad hair. Let's assume this new product is terrific—a real cure. But before selling this miracle cure for bad hair, this group has to convince customers, investors, employees, and vendors that fixing bad hair serves the diverse motives of these groups. For the potential customers, motives might include attracting mates, avoiding social embarrassment, impressing powerful others, and so on. For investors, legitimacy relies on whether the business partners can successfully stimulate customer demand, manage the employees involved in production and distribution, get vendors to take on the new product, and still have enough resources left over to make a profit for the investors. For future humans, some current stakeholders need to consider how to safely handle the waste products resulting from the production and use of the product. The mix of different motives can get pretty long.

This illustrates that the success or failure of new enterprises that follows from establishing legitimacy is not a simple proposition (G. Fisher et al., 2016). It's not just about "curing" bad hair. It hinges on meeting the complex of different interests in a group of parties to the organization. The hair care product is not likely to succeed unless the existing motives of most of these stakeholders are satisfied in some way. From another angle, the good intentions of groups working toward sustainability are more likely to have some effects when most other stakeholders' interests are adequately addressed (Amankwah-Amoah et al., 2019). The role of a legitimate voice for future humans is one that scientist–practitioners may be able to play effectively, given our understanding of human motivation.

This is why practicing psychologists typically spend considerable time on the front end of any intervention assessing the real and imagined expectations of clients before developing any plan of action. In practice, industrial and organizational (I/O) psychologists routinely follow a criterion-centric approach (Bartram, 2005; Jones, 2015, pp. 267–272), explicitly identifying the definitions of "successful outcomes" with both the hiring executives and many of the other stakeholders in an organization. This is usually accomplished by some combination of interviews, surveys, and focus groups, often as part of a needs assessment process (see Jones, 2015, Chapters 9 and 10, for details relevant to sustainability).

Applying this approach to support social speciation is problematic, however. Most criterion-centric interventions occur in large, established organizations with existing definitions of success and failure (Bartram, 2005; Muchinsky, 2004). These definitions were formed during the origination of the organization—part of the legitimizing process that I/O psychologists have less to do with. For the purpose of defining criteria for the success of initial social speciation, therefore, we need to step back and take a broader perspective based on evolutionary definitions of speciation. Following a criterion-centric approach (Bartram, 2005), I first consider the widespread failure of social speciation and the definitions of success that guide judgments of failure. This leads inevitably into ethical issues, mostly related to the long-term value of a social species—and its likelihood of leading us toward a sustainable future. The uniquely psychological understanding of human motivations discussed in Chapters 2 and 3 supports this discussion.

PERSIST, THRIVE, AND SUPPORT EXISTING MOTIVES

The premise here is that, even though failure and success are not opposites, insights into what will succeed often arise from an understanding of what fails. This is especially true for something—such as social speciation—that hasn't been studied much before. Because I/O psychologists haven't spent much time working directly with speciation and the initial organization formation process, definitions of success are necessarily a bit vague. As we get more data to work with, better descriptions and definitions will follow. Still, by trying to conceptually define a "failed social species," insights start to emerge, and ethical boundaries become a bit clearer.

From a biological perspective, any definition of speciation failure starts—and ends—with extinction. The amount of time a species lives relies on its ability to respond to the demands of its situation long enough to reproduce successfully. Extinction is one definition of failure for social species, too, except that the demands of the situation are precisely what is being altered by social speciation. So, this definition gets a bit circular. If a social species succeeds in moderating the effects of its situation, it may be less likely to fail, but it may no longer serve a meaningful purpose—it does not support an existing motive. By controlling the extremes, the motivation for change is reduced.

Another problem with using extinction as a definition is that social species may not have a "death." So, for example, some social species have been tried and gone dormant for a while then reappeared as highly successful forms later, under different conditions. The famous Greek city-state of Athens is sometimes held up as the first democratic experiment. Although it died within a century, this social form popped up over and again until it became an enduring social form much later. The Greeks did not have the means to moderate the demands of all the people who were excluded from their democratic process in the 4th century BCE. But other states tried later, after the abolition of slavery, growth of a middle class, invention of universal adult voting, public education, and other forms, large and small, that were needed to better realize democratic social species.

DEFINING SUCCESSFUL SOCIAL SPECIATION

This same gradual development of a social idea can be seen in some famous inventions, as well. The amount of time that Leonardo DaVinci's flying machine spent in the air was 0 seconds—at the time that he conceived it. The various social species (e.g., physicists, aeronautical engineers, mechanics, petroleum miners, aviators) required to make his time–space hack a reality were not extant in the 15th century in Italy.

The point is that any approach to moderating environmental demands can only be judged by the "how long it lives" definition when conditions provide for an attempt to realize it fully. Hence, the success of social species is not defined just by time—it is also defined by persistence. The persistence of a social species is a good first criterion for "success."

Drawing again on the mainstream biological view of success and failure, it is also instructive to think again of species (e.g., pandas and gorillas in Chapter 6) that have adapted into a narrow, isolated niche. These isolated species tend not to have retained the requisite variety of forms necessary to adapt quickly to the many, varied demands presented in an ecotone. The current struggles to retain nomadic lifeways and languages of conquered and once-isolated indigenous peoples suggest that some of the narrower lifeways of isolated peoples are some of the many casualties of social ecotones. As with biology, long-term isolation makes retention of old social species easier.

Likewise, some social species are just waiting for their chance to thrive in a new ecotone. These are the "generalist" social species, which we turn to next.

Thriving

A second way to define success is to consider whether a species thrives. In the case of a social species, its spread across societies and its ability to change and subsume other social species might be used as indicators of thriving. For example, religious organizations have been around for a long time in various forms. Despite substantial differences in fundamental beliefs about death, prayer, social worship, and separation between divine and human worlds, religions have been used to moderate human behavior

(Sosis, 2000) since prehistory. Religions have done this by addressing people's fears, social expectations, appetites for meaning, hopes, and an assortment of other common motives (Donahue, 1985; Hall et al., 2010; Sedikides & Gebauer, 2010). All these religious issues have dominated social discourse at some time and in some societies while being peripheral in others (Hall et al., 2010; Sedikides & Gebauer, 2010). But religions have generally thrived, meaning that they have existed in many forms for millennia, partly by subsuming other religions, partly by converting people from other religions, and partly by just changing in response to the demands of their host populations (Morris, 2010).

A thriving social species is, therefore, one that supports a lot of people, addresses existing motives, and competes effectively with other social species. In the case of religion, even today, scientific advances notwithstanding, some religions rely on their members' antiscientific bias (Rutjens et al., 2018).

By this definition of success, efforts to organize around issues of sustainability are likely to thrive, eventually, because the human motives related to climate change and waste are only going to grow, given current circumstances. Understanding the ethical problem that follows from this is important: Some social species only thrive when the threats they address stay bad—the old "dog catches car" quip applies here.

Existing Motive Support

This brings us to the third—and initially perhaps the simplest—way to define the success of a social species (or any new organization): whether the purposes of the people who founded it were accomplished. Did the roof they built stay up and keep heads dry, the heating system stay warm in cold weather, the fortress keep the enemy away, the vaccine keep more people healthy, and so on? Did the species do what it was designed to do? Did it support the motives it was designed to support? When stated in terms of the primary concern here, the speciation process needs to be directed toward some clear end, such as "staying dry," "keeping warm," "avoiding injury," "avoiding disease," and so on.

DEFINING SUCCESSFUL SOCIAL SPECIATION

Understanding this third criterion of "supporting motives" also helps to address the problem of using time as a standard for success. Short-lived attempts to realize motives—as in a flying machine that stays aloft for only 12 seconds—may not be a good basis for defining the success of a social species either. The initial failed attempts that lead to the eventual success of physical inventions are part of what defines success. The original Wright brothers plane was aloft for only 12 seconds, but there are probably flying machines over most of our heads as we read this. This means that only when an attempt has been made to test the ideas underlying a social species fully can failure be defined in terms of time. The idea of the third criterion for success might be stated as whether the aspirations—the motives underlying the social species—were eventually realized. The social speciation process may take a long time—partly because it is driven by stable human motives, and this problem of future success is particularly important when future planet mates are added to the list of stakeholders.

There are other standards for success (and failure). These three—persistence, thriving, and supporting motives—have some further shortcomings worth mentioning in the special case of social species. Unlike the usual biological definition, survival of social species is not just defined as responding to the demands of some niche. Social species are hacking the usual concept of adaptation to a niche, so even "persistence" is a bit more complicated than our definition allows. Add to this that our adaptive strategy (in the biological sense) makes humans into broad "generalists." Rather than relying on a narrow strategy, such as eating only one kind of plant, some animals (including us) adapt by developing broader tastes in food. Social speciation is broader still, giving us groups of social organizations (social species) to satisfy an enormous number of needs and wants. It allows us to shortcut some of the demands of a given niche entirely.

Nevertheless, these three constructs can provide some guidance for scientist–practitioners who are considering an active role in sustainable speciation. For example, working to change organizations that are already thriving and addressing stable needs is most of what we do now, but it makes sense as a sustainability intervention only if the organization's work is already consistent with long-term sustainability. A waste management company with limited landfill space might be looking for creative support, for example. In general, thriving organizations appear to be agile—to react quickly to change rather than being strategic—taking the long view (Pulakos et al., 2019). This may make many of them unlikely to seek help with internal speciation processes.

It may make more sense to focus efforts on helping to develop organizations that are intended to address important motives that have not yet supported a thriving species, as in the example of flying before the 20th century. Even with the relative munificence of individual life for many people in the 21st century, there are plenty of motives that are not being served particularly well. For example, organizations that have been working to profitably address homelessness, mental health, loneliness, and other unmet or underserved needs may come to thrive given a more sustainable perspective on the work they do. Persisting in what appear to be quixotic endeavors (like some of these seem to be) may be a good sign that there is a sustainable opportunity for thriving ahead.

HOPES, FEARS, AND OTHER MOTIVATIONS

As we saw in Chapters 2 and 3, social speciation relies on framing things in terms of opportunities rather than threats. At the same time, social species do arise to address fears as well as hopes, but this is when the threats in a situation are not so acute that they completely paralyze creative thinking. If the society where speciation is occurring becomes unsafe, experiences life-threatening scarcities, and loses creative capacity, failure will almost certainly befall social species that have not found ways to moderate these threats first. Where the threat is real but appears manageable, new social species are more likely to emerge.

In fact, some of the greatest inventions were designed to prevent awful events from happening—diseases, wars, and natural disasters—what historian Ian Morris (2010) referred to as "the usual suspects." Early civilizations figured out a few ways to moderate the effects of floods, for example, in years when there were no floods. This may appear obvious, but it is probably a driver of social speciation processes to have the time to think things through; still, success relies on realistically confronting some threat too. Highly contented humans have fewer motives to address. Stressed humans are busy avoiding risks in the moment. Chapter 6 introduced the idea of a "not too risky" situation and specifically social situations that have "enough conflict." As with conflict, managing the speciation process requires balancing stress with safety.

One way to directly manage this balancing of stress and calm is to alter people's perceptions of risk and opportunity. Advertisers, politicians, managers, and many others manipulate perceptions of risk, making things appear more (or less) risky or more (or less) munificent to incite action (Fernando et al., 2018; Gregersen et al., 2020; Herzenstein & Posavac, 2019; Liang et al., 2021; Skitka et al., 2017). Lower anxiety and willingness to take risks also facilitate creativity (Baas et al., 2016), and reciprocally, creative activities are used to reduce stress in therapeutic settings (e.g., Archer et al., 2015).

The power of manipulating perceptions of risk is amply illustrated by considering how a demagogue can "succeed" even in a munificent society. Recently, a large block of U.S. citizens cast their votes in response to what some would call fearmongering. A group of political candidates stimulated perceptions of serious threat at large rallies of (on average) less-educated voters—people who see changes as harder to cope with, for various reasons real and perceived. This "threat framing" in crowded gatherings led to considerable electoral success but certainly did not lead to any fresh new perspectives—never mind the stimulation of social invention.

But was the United States really suffering serious threats? Until the COVID-19 pandemic took hold, the answer was a resounding "No!" By virtually every measure, the "situation" factors in Exhibit 7.1 were ripe for creating new social species. The United States was an unrivaled military power, with an extremely robust and stable economy, high employment rates, and a high average standard of living (Morris, 2010). Prospects for continued success have rarely been better historically, and there were almost no life-threatening scarcities—certainly none that pervaded the whole society in the short term. Yet, by framing the situation in terms of threats—rather than opportunities—and relying on large social gatherings, demagoguery led to a lot of poorly managed political conflict.

We shouldn't discount the possibility that opportunity framing might lead to the failure of social species, either. Utopian notions, apparently based on visions of a better life, have turned into a number of social experiments. A common explanation for the short life of some of these communities is that ordinary human motivations were not well integrated into their creation (Molnar, 1967). But recent laboratory research shows that utopian ideals motivate people toward opportunity (rather than risk) framing (Skitka et al., 2017), suggesting that they may satisfy more enduring human motives.

This argument may fail to carefully consider a couple of key points, however. First, even though utopian fantasies appear to be framed as "opportunities," in reality, some utopian communities were also attempts to escape a threatening social world—to avoid risk. Second, successful social species can rely on both threat and opportunity framing. The Latter-day Saints appear to have framed their early appeal in both ways—trying to avoid persecution in settled areas, as well as seeking socially supported opportunities in a frontier. Still, there are a number of utopian communities that have thrived. In fact, the successful political and social revolutions of the past few centuries were guided partly by utopian road maps. James Harrington's (1656) utopian ideal of a republic of *Oceana*, for example, described an entire nation based on the election of rotating public officials—many years before this was tried in modern nation-states.

These examples point to an important way to influence perceptions of risk: social support (again). Having other people who appear to be willing to address risks together makes a difference in whether people avoid or embrace challenges and opportunities in their situations (Desivilya et al., 2010). Having a few friends around goes a long way to calming us down and hatching creativity.

MANAGED GROUP PROCESSES: SHARED LANGUAGE, SHARED GOALS, AND BEHAVIORAL FOCUS

These examples suggest several strategies for managing the process of social speciation. As a reminder (Exhibit 7.1), the processes being managed include social support for trying out new approaches to adaptation, as well as managing conflict with other social species in the ecotone. Groups that interact without such support are more likely to wind up in conflict that drives them back into narrow, time-honored, but often unsuccessful approaches to dealing with problems (Ask & Granhag, 2007; Gladstein & Reilly, 1985; M. H. Tsai & Young, 2010).

As we have seen, supports for conflict management include a shared language, perceived shared goals, a focus on behavior, and the facilitation of interactions. The modern world has some great affordances for dealing with the first of these—a shared language. Anyone with a smartphone has the ability to communicate across all kinds of distances, languages, and cultures. We can even observe others' responses on a smartphone screen to see what emotions are elicited during interactions around the globe. Clearly, there are still linguistic gaps, but from a historical perspective, language differences today are comparatively minor impediments to cooperation.

With that said, from the psychological perspective, it seems important to consider language when starting into social ecotone management. Unfortunately, there is not much applied psychology research dealing with the establishment of new languages in change management—never mind social ecotone management (Pinchot, 2021). There are certainly a host of buzzwords, dog whistles, and "insider" words used in various areas of practice, but these often change over time and lack clear shared meanings. So, although we have many advantages, the inability to understand the subtleties of language, lack of open communication between appropriate partners, and lying can derail speciation. This is why mediation often starts by establishing some shared definitions, and ground rules with specific consequences are agreed on in advance of interaction (Sturm, 2009). These may prove valuable for scientist–practitioners supporting speciation processes.

Crucially, psychologists have methods to help groups recognize the motives underlying each other's behaviors (Grutterink et al., 2013; Jiang et al., 2013; Leslie et al., 2020). Understanding another group's point of view can profoundly affect the success of conflict management. Social ecotones also provide daily opportunities for people to recognize the realities (and perceptions of reality) of the people around them. Finding ways to encourage such insight may lead to developmental change occasionally. Even without such profound change, identifying shared motives with those who appear different is certainly an essential driver of action (Dong et al., 2020; Grutterink et al., 2013; Leslie et al., 2020; Lines et al., 2021; Tjosvold, 2008).

Given grounding in some common language, though, applied psychology provides both knowledge and methods for managing the other three supports for speciation. First, we rely on an initial assessment of stakeholder perceptions of the situation. Again, this assessment usually starts with careful clarification of the interests of all stakeholders (Jones, 2015, pp. 268–270)—including, in sustainability enterprises, the interests of future planet mates. When done skillfully, such an assessment uses stakeholders' own descriptions of their interests, making both actual and perceived motives explicit for every group. These assessments also rely on structured methods designed to focus on behaviors, facilitate interactions, and reduce concerns about bias, which may explain why they are so broadly applied (Gittleman et al., 2010; Guan et al., 2011; Thornton & Rupp, 2006; Zemke, 1994).

Assessment and Facilitation

It's hard to understate the importance of this assessment process. It provides a realistic starting point on a sort of "motivational map." It uncovers definitions of success and failure, as defined by stakeholders. It identifies common ground where there is some common ground—and there almost always is, in my experience. The process allows for consideration, in advance, of how to frame and otherwise communicate the social speciation process to stakeholders. Often, a common language—such as the four letter "types" seen in some common assessment methods—emerges from discussions of perceived interests and definitions of failure and success. It provides a developmental perspective to those who are able to grasp this perspective and formulates the problem in such a way that people in earlier stages can structure their thinking. It reduces the uncertainty and risk perceptions that go with uncertainty. It provides tangible social support from an "expert" psychologist. Not surprisingly, then, the information from this initial assessment can be used to control what proceeds and what is thwarted and generally rein in or unleash various social processes.

The assessment process also provides a method that probably doesn't cross people's minds when they think of "facilitation." Averting impending failures of sustainable social speciation may require a sort of wake-up call. The sometimes-personal feedback from the assessment process can show group members how their behaviors are thwarting their efforts to persist and thrive and can create the internal conflict necessary for change. This is sometimes referred to as "holding up the mirror" to people involved in struggling groups. A feedback report from the assessment is carefully crafted, then shared with group members individually. An in-person discussion with each group member ensues after a brief cooling-off period, and then the group has a chance to share and discuss what they found.

However, even skillful application of assessment and feedback can backfire. Identifying differences and "problems" may become counterproductive, turning from problem solving to gripe sessions or reifying the differences between people identified in popular personality-based assessments. Perhaps most problematic is that facilitated interactions may help ecotone management in the short term while ignoring and jeopardizing the long-term viability of social speciation. In fact, failure often follows from sole reliance on short-term solutions (Mitzinneck & Besharov, 2019).

The Motives That Matter

As these formal and informal support functions are described, it becomes obvious what kinds of risks and opportunities drive human motivation. It is no longer predators in the wild or fear of freezing in the winter that consume our daily activities. We may never know for sure about prehistoric human motivations, but social motives are certainly essential today. When we think about risky outcomes that we'd like to avoid, high on the list are situations such as being ostracized from families and workgroups, dumped by friends, divorced from mates, abandoned by children, and other social anxieties. Likewise, when thinking of opportunities, we may still think of mocha-chocolate cake, but finding like-minded friends, falling in love and making a family, advancing in work roles, and reestablishing connections with dear old friends stand apart as the motives that drive much of human activity.

Social motives are often very enduring. Going back to infancy, we tried everything we could to get what we needed from all-powerful authorities—our caregivers. Starting in the preteen years, we tried to find love and connection through conformity. For some adults, education, travel, and inquiry rely on breaking with local convention and lead to the discovery of the treasures and trash of the wider world with our mates and offspring. Social discovery keeps some of us active in ways that nothing else can.

All of this points out what should be quite obvious by now: Human motivations are profoundly social. We don't just "need social support" to learn, develop, and act. We are driven by a fundamental need for support, inclusion, and engrossment with others. If we want to manage social speciation, these social motives are like the essential elements required for physical enterprises—metal ores for steel, microbial functions for medicine, ceramics for computer chips, and so on. Social identities are the outer accoutrement, but they belie this fundamental need to belong.

Even more important to the success of sustainable social speciation is that these motives are themselves sustaining motives. We eat cake and feel happy right away. But finding the love of one's life may take the better part of a lifetime and, once found, sustains action powerfully over an entire adult life.

There are many longer-term efforts to manage social speciation following from these sustained social motives. Several recent *American Psychologist* articles proposed a list of methods for managing some of the factors in Exhibit 7.1, most of which are not quick-fix approaches (e.g.,

Clayton et al., 2016; Collins et al., 2018; Converse et al., 2021; Gifford, 2011; Nielsen et al., 2021). Supporting community efforts to promote safety and prosperity are high on the list, especially for marginalized people for whom these are greater problems (Nweke et al., 2011). Creating manageable conflict in policy-making groups and educating the people who manage these groups' direction are strategies that often follow from the assessment and feedback method described earlier. Even in the longer term, finding ways to enlist parents and educators to encourage student self-confidence and curiosity and finding ways to encourage development change in people in leadership groups are likely to increase chances for sustainability. All take the long view. And when carefully informed by psychological science, these approaches may bootstrap creative speciation toward a sustainable future.

But risk perception remains a constant threat to these efforts. The perception of risk is easy to stimulate even in the safest of modern societies. Certainly, conflict that follows from risk avoidance has been the death of all kinds of extant social forms (Leidner et al., 2013). This is why trying to manage conflict is a key ingredient in the creation of sustainable social forms. Large international entities such as the United Nations and World Bank attempt to adjudicate or resolve the root causes for conflict as a way to lay the groundwork for such things as microloans, health initiatives, and other creative social speciation. At the local level, conflict resolution has been managed through adjudication processes such as family courts, mandated mediation and arbitration, and town hall meetings, occasionally with input from psychologists. All these have the advantages that come from taking a longer-term perspective—accepting the likelihood that efforts will sometimes fail in the short term but still have a chance of succeeding if we persist.

Interventions to Manage Risk Perceptions

Successful ecotone management also arises from something that is not so often the object of change in I/O psychology. Many people have "implicit theories" that block change (Burnette et al., 2013). Some of us assume that we are incapable of meaningful change—something that psychologists know is patently false for many aspects of human thinking. But this belief keeps people who hold it from even trying to change. Finding ways to increase self-efficacy—specific beliefs about one's ability to do something—can be an important lever in efforts to press people with such implicit theories toward creativity and sustainable speciation (Bandura, 2012).

Clinical psychologists often use behavioral practice (i.e., role-playing) to enhance their clients' beliefs about their ability to manage difficult interpersonal relationships (Kazantzis et al., 2018). For example, the clinician plays the role of a troublesome spouse, showing how the client's reactions to the spouse's aggression can change the spouse's behaviors. The client is then given "homework" to try this in the relationship. Done skillfully, this sets up circumstances where people are rewarded for curiosity testing—seeing whether their cognitive maps are correct in a real-world setting. Clinical research has shown the positive effects of helping people recognize how a change in their behavior can make situations work out differently over time (Kazantzis et al., 2018).

This is another example of facilitated conflict resolution, as well. The same self-efficacy boost can also occur without facilitation through various kinds of social support. There are many examples of coalitions and mutual aid groups demonstrating effective advocacy in public venues. These social species have succeeded in creating more sustainable futures, often in the face of considerable conflict (Riemer & Harré, 2017). Again, these efforts usually require considerable persistence.

Following from the research we visited in Chapter 7, there is also evidence that "role definition" may help some people work through the social speciation process. Being clear about who does what may be quite difficult before a social species has fully formed, but having conversations about roles as they emerge may help reduce conflict and readjust the social speciation as it occurs (Grutterink et al., 2013; Jones & Lindley, 1998; Leslie et al., 2020; Lines et al., 2021). In any case, it is one potential way to "work around" implicit theories during the speciation process.

Like most psychological characteristics, people's individual beliefs about their ability to deal with conflict probably come from earlier life experiences. This is why people who have been required to deal successfully with more conflict in their lives may be better equipped to support and facilitate these efforts. On average, people from minority groups may have this "advantage" to a greater extent than people from more privileged groups (Olson-Buchanan et al., 1998). Certainly, if social ecotones are where new social forms occur, it makes sense to listen to people who live in social margins.

ETHICAL QUESTIONS

Successful new social species rely on the three factors we have already visited: a shared language, perceived shared goals, and a focus on behavior. Once deliberate speciators have identified an existing but unmet need, they can use a common language and perceived shared goals to call attention to it, promising to address it in a new way. This promise of fulfillment can be enormously problematic for a number of reasons. In the best case, the approach to fulfillment succeeds in alleviating some risk, delivering some opportunity, or both.

The ability of a social species to fulfill its promise is a central ethical question. What happens if, instead of improving the lot of people who buy into a species, their situation is exacerbated by the approach taken by the social species? Worse, some social species may rely on exacerbating waste and destruction. For example, the disposable plastic bag is one of many examples of successful inventions that rely on exacerbating a problem. The production of plastic storage bags remains profitable partly because prosperous societies produce more food than we consume. So, it would be in the interest of plastic storage bag manufacturers to find ways to exacerbate the overproduction of food in the home. Such exacerbation of needs is just one of the ways that social species survive, but it is one that deserves the attention of people interested in reducing plastic bag use. It is also illustrative of the need for applied scientists to take the long view as we work with corporate and entrepreneurial clients.

It is possible, however, to last a long time but not thrive or meet an original need. Religions are particularly good examples of an old social species that has "succeeded" only in the longevity criterion but not thriving or dealing with human motives in a novel way. First, religions have had to redefine their purposes quite often (Frankopan, 2017), sometimes claiming to help with human suffering, sometimes advancing a "godly world" of the future, occasionally leading believers to try to pray for the end of the world (Festinger et al., 1964), along with a host of other purposes.

This may be the result of the authoritarian and conventional bases for some religious organizations (Hall et al., 2010; Sedikides & Gebauer, 2010). These may be good for maintaining an existing social species (in the case of religion, it might be better to call them a social genus or even family, following the phylum taxonomy). Monarchies share this tendency toward maintenance of conformative social species but appear less likely to incite social evolution. Because so many humans are authoritarian—especially before the widespread availability of education that stressed critical thinking—some social organizations grew up in challenging ecotones (e.g., after the fall of Rome, in the Spice Routes after the fall of Persia). Such social species certainly served important human needs during such troubled times, but their persistence in a more munificent social world remains an important ethical issue.

Although some of these long-lasting species have value propositions that attempt to enhance the lives of people affected by them or who are part of their social identity, they do not always succeed. Human enterprises are doomed to make mistakes, especially if they are founded on human motives related to risk. By this measure, they may have met some "success," having lasted a long time, but not by successfully managing human needs.

Successful but Evil Social Species

There are some social inventions that, while they may have advanced some individuals' comfort and safety, have done so at the expense of other people. Trying to understand how slavery, eugenics and genocide, survivalists, war cultures, and some kinds of bureaucracy came into being is a long and difficult discussion. It is worth considering to avoid these same massively unsustainable approaches in our future efforts.

DEFINING SUCCESSFUL SOCIAL SPECIATION

Because of its persistence through human history, its incredibly destructive influence, and its enormous economic importance, slavery is perhaps the worst social species ever to have been invented. Make no mistake: Slavery was essential to our advancement as a species. It served an essential purpose in realizing our hack. Speculating about whether we could have done things differently gets us nowhere in the face of this fact. But trying to understand how and why slavery came into being provides important insights, especially about how we avoid its long tendrils moving forward. Also, careful analysis shows how reparation—a new social organizing principle—flows quite sensibly from the origins of enslavement.

Slavery relied on the total, violent, coercive, and lethal control of one group of humans by another, usually over the entire course of their lives. In many instances, this control was exerted over generations and took on all sorts of different forms of cruelty and depravity. It has no direct parallel in the animal kingdom, which makes it a uniquely human social species that, not surprisingly, happened where human groups with different social identities came into contact. One group enslaved the other, usually after warfare, then relied on the enslaved group's labor to bolster the enslavers' comfort and well-being.

Like other novel social species, it occurred in long-lasting social ecotones, met the existing need for scarce labor, and involved conflict that the enslaving group saw as winnable. When we look at the other factors in Exhibit 7.1 for a moment, we see that enslavement also occurred when "explorers" and "conquerors" from the enslaving group saw an opportunity and had the leisure and resources needed to capture, sell, and keep people in chains—usually involving remote agricultural and mining enterprises. The social resources available to the enslavers also made this possible, particularly the absence of sanctions outlawing this social "arrangement." It required the cooperation of other powerful people from the enslaved group, who received considerable value—of various sorts—from sending their fellows away forever. New forms of communication arose to address the realities of enslavement, as well, both within and between the two groups. Sadly, slavery has all the hallmarks of a successful—and originally novel—social species.

Even from a developmental perspective, enslavement follows an expected pattern. Although we do not know the original thinking that led to it, clearly, enslavement in the West suppressed later stages of development in both enslaved and slave-holding groups through such means as laws against literacy and the discovery of other social arrangements that often accompany literacy. All forms of social interaction relied on authority and convention, with ever-present risk and shortages leaving little room for creative thought.

Perhaps most strange is the shutdown of the decision processes in Exhibit 7.1. Again, relying on historical accounts of enslavement in the American South, people are reported to have ignored enslaved people, denied any problem with the institution of slavery, relied on convention and authority—slavery as a "way of life"—instead of relying on deliberation in decisions about the social and economic order, and devalued learning and development in institutions that supported this economic and social system. Although there were concerted efforts to change this system (e.g., abolitionists, the Underground Railroad), it took violent intervention on the part of states that did not have slavery to eliminate this pernicious social species.

For the most part, this particular pernicious social species has been all but eliminated. Enslavement still occurs but is illegal and includes a tiny proportion of human populations compared with its historical prevalence. Because there are so many of us now, it is also hard to find a place on the planet where some profitable enterprise can be carried out in the physical isolation that was required for enslavement. Even in the frontier of North America, enslaved people and some of their free neighbors organized another new social species, creating the Underground Railroad to escape enslavement.

Because of and despite its cruel approach, enslavement lasted a long time. We still struggle today with "ethnic cleansing," terrorist groups, and other social species that rely on authoritarian behaviors. Keeping an eye on the conditions that lead to successful speciation may help avoid these problems in the new social species which have sprung from these older ones. Certainly, scientist–practitioners trying to steer speciation toward a sustainable future need to keep the factors that spawned enslavement from taking root in new species, however desperate sustainability issues may become.

Unethical Tactics

Some persistent, thriving social species meet needs that support a more sustainable future but use tactics that rely on what might be considered destructive or at least unethical. Most people would agree that telephones, including the modern minicomputers we still call phones, have met real human needs, including serving sustainable ends. We don't need to saddle up a horse or rev a car engine to visit with friends and family, go to market, and so forth—phones have reduced the human carbon footprint in this way. However, the tactics used both in the introduction of hardware and the uses of software have crossed some ethical lines. The withholding of new technologies has cost time, money, and other valuable resources for millions of people—at the same time enriching large corporate entities (Dunford, 1987). Here's how it works. Technology such as cellular transmission is invented by a small startup company. This company is bought by a larger company, including the patents for the cellular technology. Although immediate production and dissemination of the new technology would save customers time, money, trouble, and so forth, the company wants to soak up all profits available from the old technology they invested in and released earlier. So, to maximize profits, they wait for the profit curve to start flattening before releasing the new technology.

This approach takes the interest of profit ahead of other important public goods. It also demonstrates how, in so doing, new social species are barely given a chance to survive—never mind thrive or provide for some affirmative motive. But there appear to be no restraints on practices like this and others that give advantages to existing social species and kill new speciation. There may be ways for applied psychologists to weigh in on decisions by larger organizations, allowing new species to thrive as separate but "owned" entities for some period in order to consider their efficacy and long-term profitability.

When deciding how to use the psychological avenues described here in pursuit of some important action, I use a straightforward rule: Never consult alone. There is some evidence in group decision-making research that ethical choices are more likely to be made when people collaborate with respected, knowledgeable colleagues willing to take on the devil's advocate role. When in doubt, ask a trusted colleague for counterarguments and ways to manage contingencies. To be clear, you will still make errors, but at least you will have a better idea of why and have some plans in place to correct the situation.

SUMMARY

Getting a diverse group of people who are high in agency—whose basic needs are met and who are developmentally advanced—to share some common purpose and constructively manage some common conflicts can lead to some wonderful new social species. Business and governmental forms, professions, philanthropies, community mutual aid groups, and more have all arisen under the right circumstances. But these circumstances also can lead to some horrors.

We have seen in this chapter that the conditions that support speciation processes are already managed in many work organizations, communities, and even nation-states to meet the three criteria for success—persisting, thriving, and serving existing motives. It is also clear that, even though they are interrelated, it is still possible to meet one of these three criteria but not the other two. So, although persisting is required for thriving in the short term, it is possible for a species to go dormant for hundreds of years, waiting until the necessary conditions are in place for it to later thrive. It's also possible to meet some motive without necessarily thriving and thrive without supporting a consistent motive. These three criteria likely have complex interrelationships, but they still serve as sensible standards for judging whether species management efforts are "succeeding."

Regardless of success or failure by these three criteria, there will always be concern over both the motivational purpose and the tactics used in achieving "success." Slavery and totalitarianism serve some of the darkest hopes and aversions in our motivational repertoire. Likewise, some of the highest hopes and most realistic threats we face are served ruthlessly—without considering the long-term consequences of choices. Sustainability relies on the consistent inclusion of "outsider" voices from an unknown future as these choices are made in all sorts of organizations.

Clearly, complex social species that rely on careful invention and reinvention have proliferated rapidly in the last 3 centuries. The revolutions—some violent and some not—during this period have yielded an increasing number and variety of social species but also social species that rely heavily on an educated, developed population for their invention and maintenance. Representative forms of government may have existed in small groups in the distant past, but only in the past few centuries have nation-states been able to make this approach work in much larger populations, with many more mixed social identities living in close proximity. It is not a coincidence that the levels of literacy, years of compulsory education, and an increase in intellectual capital for human enterprise have been accompanied by this boom. Looking at the modern political landscape demonstrates this, as well, with nationalist leanings—based on authority and convention—relying heavily on less-educated, generally less well-informed voters (Carvacho et al., 2013; Cuevas & Dawson, 2021). Working toward a shared mental model in a large enough majority of the human population is an overarching aspiration for a sustainable world.

9

Taking Action

Few scientists seem to have grasped just how essential social speciation is to the success of *Homo sapiens*, much less that there is a scientific literature that provides evidence to help manage the speciation process. Perhaps this is because it seems so obvious to biological scientists that social organizing has been necessary to human survival. But it also may be that so few of the applied psychologists engrossed in social organizing have considered the long-term effects of unregulated social speciation. Fortunately, scientist–practitioners have a substantial scientific literature supporting efforts to manage families and workplaces. Hopefully, the perspective on social speciation processes offered in this book will prove valuable to the pioneers who have begun applying scientific psychology to sustainable organizing (Amel et al., 2009; Gifford, 2011; Klein & Huffman, 2013; Ones & Dilchert, 2013), following Rachel Carson's (1962) lead.

MANAGING OUR HACK

On a grand scale, social hacking has created planet-wide, catastrophic changes that have been identified by the social hacking method we call science. Having the ability to test both physical and social contingency maps, humans have moderated the environmental pressures that lead to genetic evolution and behavioral learning. But we are only starting to use science to manage ourselves or, more precisely, "manage our hack." It is past time to apply what we know about social organizing to try to manage this process.

Assuming that social speciation processes occur in social ecotones, where different social systems and social identities come into contact, there is a lot that needs to be done—and the sooner, the better. As different human populations have expanded and come into contact with one another, different social systems have interacted more often, sometimes leading to greater conflict, developmental change, and, recently, rapid social speciation. As populations continue to interact at greater rates in relatively safe situations and before the massive tolls of planet-wide change start to make things less safe, the need for coordinated management of this speciation process has become urgent.

Applied psychology provides a starting place. We have a partial map of the conditions under which innovations are likely to occur in social margins. Given this understanding, what sorts of action are likely to help us manage such speciation and help or hinder our long-term survival?

Preentry Preparation

If we treat social speciation as a process, the finer points of the map start to reveal themselves. Exhibit 9.1 is a preliminary description of this process, including some of the psychological interventions likely to affect sustainability at each step. The great challenge is to find lasting, flexible, effective ways to direct these processes toward sustainable futures. Exhibit 9.1 is an attempt to organize thinking about how this might work, using applied psychological knowledge and methods.

Exhibit 9.1

Managing the Steps in the Social Speciation Process

Step 1: Preparation for Entry Into the Ecotone

Crucial member roles: Boundary spanner, futurist, entrepreneur

Psychological variables: Openness, empiricism, risk-taking ability; Developmental Stage 3

Possible interventions, training: Realistic preview, conflict-management skills, brainstorming and devil's advocacy, self-efficacy bolsters, assessment of group similarities and differences, appropriate humor

Step 2: Entry Into the Ecotone

Crucial member roles: Environment scanner, intelligence gatherer, interpreter

Psychological variables: Perception of risk, social anxiety, tough-mindedness

Possible interventions, training: Stakeholder analysis, strategic planning, stepladder technique, cultural awareness and empathy experiences, intelligence reports, conflict management, prejudice reduction

Step 3: Contact and Interaction

Crucial member roles: Initiator, communicator, entrepreneur, conciliator, creative thinker

Psychological variables: Openness, linguistic skillfulness, agreeableness, humor; Developmental Stage 3; comparisons between groups on these variables

Possible interventions, training: Training in relevant cultures, reflexivity interventions, conflict-management skills, recognition of motives and expertise, devil's advocacy, appropriate humor

(continues)

> **Exhibit 9.1**
>
> **Managing the Steps in the Social Speciation Process (*Continued*)**
>
> **Step 4: Conflict, Negotiation, and Creation**
>
> Crucial member roles: Inventor of language, interpreter
> Psychological variables: Tough-mindedness, risk-taking ability
> Possible interventions, training: Same as Steps 2 and 3
>
> **Step 5: Building a New Social Species**
>
> Crucial member roles: Manager, interpreter, conciliator
> Psychological variables: Openness, empiricism, risk-taking ability
> Possible interventions, training: Realistic preview, conflict-management skills, executive skills development, long-term perspective

First, if we know in advance that different social species are likely to come into contact, several approaches can be taken that will affect later decisions. Psychologists might assist in deciding whether aspects of these impending ecotones should be encouraged, disrupted, blocked, or otherwise manipulated in advance of contact. Likewise, identifying which groups to engage in advance and preparing them to deal constructively with likely conflicts may help direct interactions toward sustainable speciation (see Nweke et al., 2011). Psychologists also have tools that help assess and identify the specific circumstances in the groups, such as whether they are developmentally prepared, perceive things as risky or opportune, are ready for conflict, and so on. This may help set things up so that when they do come into contact, people may understand that changing their cognitive maps is likely to be necessary for the successful adaptation of their group (B. Koh & Leung, 2019). Perhaps most important, preparation could help groups consider and define their roles, including who is ready to assume leadership under what circumstances.

Psychologists from several applied areas have engaged in these activities already. Examples include selecting people for self-managing groups

(Jones et al., 2000), creating and facilitating decision-making and brainstorming groups (Rogelberg et al., 2002), bringing different stakeholders together to solve community problems (Collins et al., 2018; Mitzinneck & Besharov, 2019), and designing conflict resolution procedures (Čehajić-Clancy & Bilewicz, 2021; Olson-Buchanan et al., 1998) before deliberate creations of social ecotones (e.g., organizational mergers and corporate acquisitions). Their success has been determined partly by the integration of research, theory, and ethics into application.

This preparation phase may also be a particularly fertile time for interjecting the "future stakeholder" voice in discussions. Research on this "timing" question may prove pivotal.

Managing Transitions in New Ecotones

Once new ecotones have occurred, psychologists can apply a number of successful interaction management methods. In fact, psychologists have taken leadership roles in everything from corporate mergers (Woehler et al., 2021) to providing services to migrants and refugees (Thompson et al., 2018). Taking on the voice of the "future" minority may require some of the most difficult challenges for sustainability. For example, should decision-making helpers (scientist–practitioners) ever take on the "consistent and relatively extreme" voice of a minority to elicit normative change? Should a member of the decision-making group be assigned this role? At what point in the process should this voice be heard? These are questions worthy of further research.

For those who have had a hand in creating an ecotone, there is some ethical obligation to help moderate and guide the behaviors triggered by it. Exhibit 9.1 provides suggestions for some active roles that may be needed here and the kinds of support and facilitation applied psychologists can provide. From the time people enter a new social setting onward, psychologists can help facilitate initial entry and contact, conflict and negotiation processes, and decisions that lead to deliberate speciation. All these can be influential in the development of a species behaviors, values, and norms. Psychologists have knowledge and tools to help manage each of these steps.

Supporting the Consolidation of a New Species

There are recent examples of new social species that are worth mentioning, partly because some have started to rely on psychology for their realization and partly because they are set up as sustainable enterprises. One is the urban farm. This new species turns the ancient "town market" on its head. Instead of producing food in a broad countryside, then transporting it to town for sale and trade, a growing number of urban farms produce food immediately adjacent to the central residential areas where it is consumed. The roads connecting the countryside with town rely on enormous amounts of fossil fuel just to transport food, never mind the cars driven to the markets from outlying residential sprawl.

Successful urban farms tend to follow old "for-profit" and "co-op" models, sometimes to their detriment. Idealistic (mostly young) people are blended with farm members looking for a "safe" social place. Psychology has dealt with such relationship issues in everything from family practice to executive development interventions. Exhibit 9.1 provides a few ideas for roles of applied psychologists in the process of developing these ecotone-spanning organizations. Applied psychologists could use the social speciation perspective to help organizers of urban farms manage choices about criteria and stakeholder motives (Akemu et al., 2016), expertise-based roles (Lines et al., 2021), risk perceptions, and management of local ecotone conditions. Similarly, applied psychologists can assist in preparing for the realities of conflict and negotiation (Mitzinneck & Besharov, 2019), help to formulate "hybrid social identities" (Cornelissen et al., 2021), and even help devise shared mental models as aspirational maps (B. Koh & Leung, 2019).

Unmanaged Speciation

This points to the reality that, up until now, psychology has often played a peripheral role in initial speciation. Sometimes everything has worked out well. In particular, liberal arts education is a famously successful ecotone. Educational psychologists have played important roles in designing and managing entry into liberal arts colleges, and psychologists from almost all subfields deliver content across disciplines, including business schools, natural and applied sciences, the arts and humanities, and of course, social sciences. We train personnel and provide staff for students from pre-entry through matriculation, serving as admissions personnel, advisers and counselors, expert speakers, recommenders, and job placement personnel. But with a few exceptions, and for whatever reason, psychologists have rarely taken leadership roles, providing support for decisions about issues that undergird the liberal arts mission of sparking developmental change and directing a successful transition to a productive and fulfilling adulthood. Despite the reliance of these sorts of decisions—as well as decisions about funding, stakeholder engagement, intergroup conflicts, and others—on psychological research, we have not often taken leadership roles in these conversations.

Because of our applicable insights, scientist–practitioners need to make this leap to take on such leadership roles in the kinds of educational and institutional reorganizations that can foster social speciation. Following the same example, liberal arts colleges and universities have spawned many prominent social species. Recent examples include Microsoft and Facebook, both of which were conceived by brilliant undergraduate students while they were experimenting with cognitive maps at Harvard University. Both are now among the largest, most innovative corporations in existence (Church et al., 1998).

But here, the similarity ends. Microsoft has succeeded in creating software to support all kinds of activities, some of which weren't thought of before they were invented by Microsoft scientists and engineers. Facebook is an ecotone by its very design. It has suffered occasionally from a sort of Wild West free-for-all, where ill-considered, occasionally malevolent social groups have incited problems in broader society (LaFrance, 2020). Facebook has even been criticized for using practices devised by psychologists, but which appear to rely less on scientist–practitioner ethics (Lewis & Wong, 2018). More ethical standards have now been instituted, including an oversight group. Like Microsoft, Facebook has employed scientist–practitioners since early in its existence (J. McHenry, personal communication, September 3, 2021), but the problem of having their voices heard will remain an important one for sustainable decision making.

As a thought experiment, what would have happened if Harvard had employed applied psychologists with a portfolio that included advocating for sustainability to support the start-up ideas of Bill Gates and Mark Zuckerberg? Might Harvard have seen Gates—who has expressed consistent and sincere support for sustainability issues (Wenzel, 2021)—all the way through to graduation? Might Zuckerberg have led Facebook toward the kind of explicit role in conflict management that would have made its novel, deliberately constructed ecotone more likely to support sustainable speciation?

The answers are, of course, unknowable, but the point remains: Applied psychology needs to find a place at the table in start-ups, including higher education, think tanks, business incubators, and other venues where initial speciation processes are happening. In addition to increasing chances for a sustainable human future, the more immediate effect of this would be to widen the engagement of applied psychology in important creative decision making (D. Liu et al., 2016) at an earlier stage of speciation, where future human needs can be addressed, along with those of current stakeholders.

Arguably, though, deliberate ecotones are what have spawned some of the great changes of the past half-century. New social species that deliberately apply psychology have stimulated financial and community support (e.g., Doctors Without Borders; Hilscher, 2016), advanced controversial human rights (e.g., LGBTQ rights; T. Thomas & Panchuk, 2009), and rapidly adapted to complex, large-scale catastrophes (e.g., some nations' responses to the COVID-19 pandemic; Domínguez et al., 2020) similar to the sort that result from climate change. Moving forward, quick, effective processes for developing shared mental models (Lines et al., 2021; Schippers et al., 2013) may prove particularly vital.

Creating an Ecotone

Using psychological applications to establish social ecotones may provide our best pathway to a sustainable future. The growing diversity of people in higher education and workforces, continued migrations, the increased number of corporate mergers, and other social cross timbers have coincided with the growth of applied psychology in all these settings, often working in the margins between social contexts. Psychological methods have supported the creation of new social identities and structures—it's just possible that some of these may help save us from extinction.

By taking leadership roles, scientist–practitioner psychologists may help construct more well-considered ecotones, effectively "hacking our hack" because it controls the conditions and processes of social speciation. Creating ecotones that support speciation will have succeeded at least in hacking our hack—even if we do not ultimately create a sustainable social world. In addition to the assessment and feedback skills we already possess, applied psychologists need to be well equipped to manage risk perceptions, highlight shared motives and otherwise manage conflict, and identify a common language to develop a shared mental model.

There is considerable reason for hope, just based on this idea of hacking our hack. More ecotones create greater diversity. Integrating another idea from biology, diversification (in this case of social forms) is the key to species survival. If conditions change dramatically, the more social species humans have at our disposal, the more likely that one or more of them will support our survival. This is another reason carefully designed ecotones may hold great promise.

We have some good ideas about the psychological conditions for social speciation, so the door is open to creative, ethically grounded efforts to form such ecotones. Other professional areas have already invented new kinds of "common" physical spaces (not the same as "public" spaces) that are safe and welcoming. Arts groups such as the yarn bombers have found ways to encourage fun, engaging, creative interactions among groups that are otherwise unlikely to interact. Surely psychologists can identify ways to do the same kind of social ecotone building in education, workplaces, virtual and live communities, co-ops, marketplaces, and perhaps some entirely new social functions and forms.

If the social speciation perspective is correct, directing the speciation process toward increasing the chances for a sustainable future will require understanding the conditions that instigate such creativity and developmental change. In particular, developing long-term perspective taking will help direct people who are engaged in ecotone management to consider future stakeholders, as well as any contingencies of current decisions for future outcomes.

SPECIATION WITHIN LARGER SOCIAL ENTITIES

It is even possible for a single psychologist to provide a sort of artificial ecotone in a large, established organization. One illustrative example of deliberate speciation within an existing social system comes from the development of a "team" work form among administrative staff in a large insurance company. Here, an industrial and organizational (I/O) psychologist teamed with a manager and worked with other stakeholders to develop a change process.

Big insurance was an early adopter of I/O psychology as a risk-management function—consistent with their core business (Dionne, 2013). Insurance companies—once called "mutual benefit" schemes—got their start with forms such as medieval artisan guilds, the 19th-century German national health scheme, and the mutual benefit pools of early 20th-century farm cooperatives (Reid, 2010; Van Leeuwen, 2016). All these forms have been successful in several regards, but all have also changed in attempts to control their social environments better. Modern insurance, which relies heavily on the statistical analysis of human behavior, is a social species that has come to exert enormous influence in recent times (Reid, 2010; Van Leeuwen, 2016).

In this example, a young and ambitious manager wanted his group to work directly with an I/O psychologist to develop an "empowered team" approach to meeting difficult organizational challenges. In particular, the top executive was working to create a market advantage for her division, mainly by relying on creative approaches that would reduce the need for the highest cost item on the company balance sheet—people. She was honest about the most difficult aspect of this initiative—implementing a reduction in the workforce. She believed that there was too many administrative staff and not enough novel approaches to dealing with administrative chores, such as answering customer questions, processing routine claims, and pricing policies.

The group's charge was to try to develop some of these more efficient approaches for the corporation, more broadly, starting with engaging line workers in the process of innovation—along with downsizing. If you think back to "risk aversion" and its relationship to cognitive learning, you will see the core problem of trying to open people who fear for their jobs to new learning and thinking. The approach taken focused on a popular term at that time—*empowerment*—which meant getting frontline workers to figure out how to downsize their own work units by devising creative new approaches. The approach chosen by the psychologist relied partly on humor—working with the young manager to reduce administrative staff's stress about job loss by perspective taking, surprise, and laughter.

Several managers openly scoffed at the idea that workers could effectively "fire themselves." But the manager believed that this approach would work if it were managed appropriately. Empowerment started with an outsider (the psychologist) asking about downsizing. This discussion reiterated the executive's honest appraisal—that this would have to happen due to external pressures. The group reactions were discussed openly, and they discussed how to fashion an approach to dealing with this deliberatively. After humor exercises, small study groups were formed around areas of expertise to answer realistic questions that the group had fashioned. The groups reported back frequently and came up with approaches they had seen before, but they also considered anything that might work.

The end point, about 6 months later, was a presentation by this work group to all 38 managers, nine directors, and the CEO that was met with unanimous approval, cheers, applause—and laughter. Furthermore, it was implemented.

What had happened? What was different here than in most human efforts in corporations? How did this group beat the assumption that people can't fire themselves? Several things stood out. First was committed, consistent leadership that kept to their promises as much as possible and dealt directly and flexibly with the realities of human motivations, conflict, and collective action. Then there was the developmental perspective that the group learned through humor—addressing fears by laughing together and developing perspective. Given their education and experience, many of the employees were likely past authoritative and conventional reasoning stages of development.

Less obvious, but perhaps most important, was the interjection of a new viewpoint into the thinking of the work group. The I/O psychologist, while addressing a conventional corporate culture, used some different but well-supported assumptions about human motivations and what criteria were likely to be achieved—in particular, a change in the shared mental model. This probably made a difference in how this group approached their problems. It created a social ecotone where a new social species (I/O psychology) blended with traditional management (the manager in charge of this group) in a munificent environment (lots of resources available for everyone—even the "losers" in the musical chairs) to create something of a new social form for this corporation—the empowered work group nested within a corporate bureaucracy.

The second, more dramatic example of this kind of psychologist-ecotone comes from another I/O psychologist: Dr. Mahima Saxena of the Illinois Institute of Technology. Dr. Saxena's academic interests lie in the study of workers in the gig economy, particularly those who are poorly paid for their work. Some of her work blended the "cutting-edge" technology of real-time behavior tracking with the ancient work done in rice paddies (Saxena, 2016). Workers were prompted by a pager to describe a few characteristics of their current situation, including such apparently mundane things as mosquito prevalence. Analyses of the data from this experience sampling technology yielded results that were anything but mundane.

The discovery that came from this meeting was that the incidence of mosquito bites increases at certain times of day, which reduces productivity by taking workers who get sick from these bites out of commission. I submit that this blending of the pager with subsistence agriculture is another example of an artificial ecotone, where two distinct social species (person-focused high-tech culture and rice agriculture) meet. The fact that it led to a restructuring of the millennia-old work schedule of crop planting and harvest makes this example matter. This adaptation, which seems like such an obvious one, is in fact revolutionary, not least because of the potential life-saving impact, along with the increase in productivity.

FORMS TO SUPPORT PROCESSES

Going back to the flying analogy: Physical structures were combined so that sometimes common behaviors (moving front limbs) led to a new outcome—flying. The analogy here is that the conditions for social speciation have led to some "same old behaviors" winding up with different outcomes. The conditions for taking a risk were in place, including social support, inadequate but not life-threatening resource shortages, and so on. But the same old behaviors that we had learned reactively somehow got humans to start building things together.

This brings us full circle. How do we use the same old motivations, abilities, and situational variables to better manage the social speciation process that has made us so successful so far? We can certainly manage the existing social artifices we have created. But what we're trying to do is to manage the same process that built these artifices to alter them and create new versions of them that will work better for a sustainable future. By focusing on the conditions and forms that guide the social speciation process, we may be able to hack our hack.

MANAGING THE OTHER FOUR FOES OF SUSTAINABILITY

All the levers in Table 1.1 are usable in support of these strategies. In fact, all have been used extensively in modern societies for everything from education to incarceration. I did not mean to entirely dismiss them as tools to support our efforts toward sustainability. Instead, it is important to see them for what they are—reactive changes. But we still need to get at least some people to attend to strangeness, see a problem when there is one, think critically about the realities of our situation, and be adept at learning in its various forms. But what we need most is proaction that relies on these and other psychological levers to manage ourselves.

We've also spent time talking about the individual characteristics that improve our chances of successful speciation. Rather than dwelling on these, it makes sense to reiterate that perceptions of situational pressures and risks can overwhelm their realities. This can cancel the effects of individual predictors entirely sometimes (Jones & Parameswaran, 2005). Furthermore, the same social conditions that spawn and destroy new social species can change individual characteristics that help or hinder social speciation. For example, parents overwhelmed by difficult situations do not encourage curiosity testing when they see the world around their children as "dangerous." Their children will not grow up to be curious in a world where there is no time for curiosity. Likewise, "relaxing" is unlikely to lead to survival when people face real existential threats. It is an uncommon person who sees themselves as able to respond creatively to a highly threatening situation—and it's not individuals who do this, anyway.

SUMMARY

Some of the tools for managing social speciation appear to be readily available—even obvious. Identifying and using critical levers for encouraging creative group mental model building is a particularly promising approach. Other approaches need to be invented, tested, and elaborated. Applying existing methods to the complex task of ecotone management may be a potentially powerful approach for this. These and other boundary-spanning efforts are waiting for dedicated scientists to test and put to good use in many venues, both public and private—so is the careful ethical discussion of the potential and actual results of these new process management attempts.

10

Solutions

Psychology provides a different framework for defining sustainability. It also provides workable tools to change thinking and behavior in real-world settings. Applied psychologists have invented many of these tools and put most to work in social settings such as organizations, families, and communities. We have done this by focusing on human social motives—being "safe" in a social space, being "included in a conforming social identity," and taking ethical leadership roles where people in earlier developmental stages are more concerned with safety and inclusion than with a viable future. Like other adaptations, applied psychologists' engagement in social speciation has often led to failures and missteps, and it has also led to ethical lapses. But also, like other adaptations, there have been some brilliant, sustainable new social species that have grown, thrived, and deliberately hacked the evolutionary system under circumstances that were guided by applied psychologists.

---

https://doi.org/10.1037/0000296-010
Sustainable Solutions: The Climate Crisis and the Psychology of Social Action, by R. G. Jones
Copyright © 2022 by the American Psychological Association. All rights reserved.

REFRAMING THE SUSTAINABILITY PROBLEM

What actions will help or hinder our survival? This is a big, burning question. To those inclined to take on the many challenges of sustainability, it is *the* question. Answering it on behalf of future life is what matters. This has led to all sorts of excited people arguing that their new technology, new government regulation, new market idea, or new manufacturing process will get us out of one of the holes we've dug. Famous authors, scientists, and politicians offer solutions they think will solve our environmental problems. There is a growing recognition that "sustainability" is more than a commercial or political buzzword—and that the "solutions" being offered don't seem to be making the difference we had hoped.

These solutions sound sadly familiar to an applied psychologist. They are almost always stated in ways that imply a psychological solution, generally without the benefit of psychological science. Some variant of one of the five popular foes of sustainability from Chapter 1 (see Table 1.1) is paraded out as "the way" to reduce our destructive human footprint. We need to "get people to attend" to the situation or acknowledge that "we have a problem" or "think more deliberately" about our choices or "educate people better" about the nature of the environment. Few call for direct social action (related to the fifth cause of unsustainable behavior). Few of these popular answers are defined using psychological science.

To her great credit, the brilliant scientist Alice Roberts (2017) suggested that sustainability is a "tangled, twisted" set of problems without any obvious or simple solutions. She made a general observation that relates to the argument I have made here—that we are extraordinary social organizers. Even to the ears of a psychologist, her insight that our adaptation has relied on our ability to "cooperate and help each other" makes sense. But because she and Noah Harari (2014)—whose work she built on—don't integrate psychology into their description of "cooperating and helping," it remains a vague reference to something in serious need of elaboration.

Now that you've read this book, I hope that you can see the problem here. Developing a new technology, adding a new regulation, developing a new entrepreneurial enterprise, engineering a new distribution process all rely on social speciation. Our unique ability to rapidly organize our social groups in new ways has gotten us into the mess we're in. Most of the popular "solutions" on offer assume that we can use this same hack that got us into trouble to get ourselves back out again. None gives a detailed description of how this hack works. The fact that cooperating and helping each other is held up as a "solution" demonstrates how uncritically we have treated the core psychological processes required for sustainability and how illiterate many are about the large body of empirical research that addresses these processes.

This raises some specific psychological questions. For one, how do we use the motivations, abilities, and situational variables that stimulate the social speciation process to manage these same processes? We're not so much interested in managing the existing social artifices that we have created. Instead, we're trying to manage the same process that built these artifices to alter and create new versions of them that will result in a sustainable future. We're sort of trying to build our way out of our building.

Reframing sustainability as a psychological issue makes this the central question: Can we better manage the social speciation process that has led to our problems as a way to solve these same problems?

BEYOND INDIVIDUAL ADAPTATION

To review, our core adaptive strategy is to build social species to support the construction of artifices that moderate the environmental pressures that are the main forces affecting genetic selection and learning. Success at moderating these pressures exempts our species from some of the forces of evolution—at least for a time. We are evolutionary hackers. From a psychological perspective, individual motives to stay safe, comfortable, and satisfied have led us to organize around shared mental maps to reduce the incidences of extremes in the world around us. People figured out (for better or worse) how to organize together and take action to realize ideas for keeping threats (e.g., cold, heat, germs, predators) away, increasing the abundance and reliability of food sources, and so on. Once one problem was solved in this social organizing fashion, it tended to get passed along, but this didn't stop us from social invention.

We have created a cycle in which deliberate alterations in the environment increase individual comfort, which creates conditions for more creative thinking and testing of creative ideas through social speciation, which leads to further altering of the environment. Rather than being subject solely to the adaptive processes of other species, changing the structures and behaviors of individual humans through DNA selection and RNA learning, we have increasingly found ways to dull the effects of these processes. Given time and a lack of close focus on survival activities, people have thought up all kinds of creative new concepts to increase individual safety and comfort—and also new ideas for organizing groups to try to realize these concepts.

The rate of social speciation has also been spurred by the tendency of peaceful contact between groups with different lifeways to increase creativity. As we have become safer and more comfortable, human populations have surged, making it more likely that different social groups are forced to intermingle. The relatively safe social ecotones created in recent times have led to further increases in the rate of social speciation. We've altered our environment bit by bit for most of our species' time on the planet. Much like other forms of evolutionary adaptation, we've been reactive in our hacks. It got really cold in the subarctic, so people organized themselves around fire pits, yurts, and igloos. But having diverse, comfortable, safe groups with different mental maps come into contact has made us even more profligate hackers—video games don't increase safety or comfort, but they sure can be fun!

There are other explanations for the rapid rate of social speciation in the past few centuries, but this cycle of comfort, creativity, social species interaction, and further speciation rests on good evidence and gives actionable targets for managing our hack.

WHAT TO TARGET FOR CHANGE

If we hope to change our socially nested individual behaviors, we need to manage the social speciation process better where and when it occurs. It makes sense to focus efforts on the conditions that lead to successful social hacking: targeting existing human motives, managing perceptions of scarcity, constructing and managing "safe" social ecotones, establishing shared forms of communication, managing conflicts between existing social species, and providing social support for testing creative cognitive maps in existing groups.

Understanding these conditions on which social speciation relies is not enough, though. If we hope to manage speciation in the interest of future planet mates, their interests need to be considered in the processes through which speciation occurs. These processes are not well understood, but Chapters 8 and 9 provide a general "map" on which we can rely for now. Understanding how to use psychological applications for steering speciation is the key to a sustainable future. This includes fomenting new species, nudging the speciation process one way or another, and continuing to try to influence the growing multitude of social species through existing disciplinary areas of applied psychology.

Some tactics for managing the speciation process have deep roots in psychological research on minority influence, leadership, humor, emotion, shared mental models, and developmental change. Applications in industrial and organizational (I/O), forensic, clinical, educational, community, and consumer psychology are available for expert use. But the application of psychology to the development and care of sustainability remains an important frontier.

APPLIED PSYCHOLOGY, PLEASE

Applied psychology is an essential lever. It has a rich history of meaningful contribution to the development of important shared mental models, including the consumer economy, targeted combat methods, open offices, drug courts, experiential education, government policy making, selection of critical personnel, and a host of other important innovations during the past half-century or so. In fact, applied psychology went from not existing at all to spawning some of the fastest-growing occupations in the world in the past century, much like engineering did in the 19th century. As of 1900, there were no clinical psychologists, no I/O psychologists, no human factors engineers, no consumer behaviorists, no sports psychologists, and, in fact, no applied psychologists at all. There are lots of us now.

Clearly, applied psychology has found its way into making us safer and more comfortable in homes, workplaces, sports, hospitals, and just about every one of the social systems where most human activities take place. It has also led to examples where individuals with advanced training in applied psychology have made differences in the social structures where they were embedded. One example is a logistics manager at a large trucking company who worked with psychologists to successfully reduce emissions by managing a cluster of truck driver behaviors.

Psychological methods have also worked well enough to have supported the creation of entire new social identities and structures under the heading of "psychology." These have borrowed from biology the idea that diversification (in this case, of social forms) is the key to species survival. The blossoming of many applications of psychology both under their own headings and in many other professional areas signal a growing realization that "managing ourselves" is a challenging but essential process. Perhaps it is time for another new psychological discipline: applied evolutionary psychology.

Short of forming this new social species, there are successful recent examples of psychology applied to social speciation. Among other things, these examples help to elucidate the newness of this as a research area. Psychologists are employed to evaluate the viability of new enterprises for venture capital firms, assess the likely success of leadership, and train incipient entrepreneurs for sustainable enterprises. Relating to sustainability, in particular, some "incubators" use guided psychological experiences for developing new, sustainable enterprises. Psychologists have been involved in writing new energy and environment policies for governments, developing strategic plans for nonprofit environmental organizations, and providing institutional assessments for transportation startups.

Perhaps a less obvious but particularly promising approach for which applied psychologists are particularly well suited is the construction of new social ecotones. Certainly, the many professional conferences and conventions spawned by psychological organizations have given us experience in constructing gatherings somewhat akin to Burning Man. It could even be fun to deliberately convene diverse, multidisciplinary groups, each asked to devise some creative illustration of their ideas for advancing sustainability. This would be followed by defining groups according to some set of criteria drawn from the illustrative presentations.

Thinking big, it makes sense for professional scientist–practitioners to lead the formation of new organizations aimed at driving policies, ventures, entire economic sectors, and even political parties toward a stronger emphasis on the psychology of sustainability. As a relatively new, flexible social species, applied psychology is well positioned to take on any number of important change efforts in the interests of future planet mates.

A PREDICTION OR TWO

Although I am not inclined to try to predict an unknown future, I do believe that the rapid increase in social invention has reached a point where we are starting to see dramatic shifts in the way we organize at the broadest levels. Nation-states are giving way to global entities for commerce, governance, and service, just as religious species have done for a couple of millennia now (starting in the late BCE years). I offer no specific predictions, but I am pretty confident that we will see quite a bit of change in how we govern ourselves and serve our groups—not so much our individual selves.

In any case, it is my firm belief that the great scientific advances of the next few generations will be in psychology and the sciences and professions that are borrowing from it. Certainly, developing and establishing sustainable human futures will at least benefit from these advances—but I suspect it will depend almost entirely on them.

I believe that some of our current efforts aren't doing what we need them to do. Focusing on melting glaciers, severe weather events, air pollution, and wars in drought-ridden countries will certainly not help with this. In fact, given what we know about the origins of creative social speciation, focusing on these risks may make things worse.

It also bears mentioning again that changing individual behavior, even on a massive scale, while it may be an end measure of success, is premature. Until we figure out how to manage our social behaviors, widespread individual behavior change is unlikely. Focusing on action (Outcome 5) instead of numbers 1 to 4 in Table 1.1 starts with finding ways to change the social species that surround individual behavior.

It sounds a bit backward to put the "take action" target ahead of the other four sustainability levers in Table 1.1, but it makes sense as the shortcomings of the other four targets become clear. Furthermore, targeting action relies on the people who are "in the trenches" making the important decisions rather than some "psychologist in charge" proclaiming changes in behaviors. Along with the ethics of stakeholder inclusion, having people managing their own choices is likely to lead to more lasting change.

OPTIMISM AND SELF-EFFICACY

Many people would like to know how and whether we will adapt and survive. This book provides a carefully constructed, compelling, but not comprehensive answer to the "how do we survive" question. Psychological research has shown that if we hope to answer the "whether we will survive" question, there is an essential first answer to the how we survive question. If we hope to survive, the conditions that lead to taking action need to support action. The motivational equation starts with the assumption that our efforts have some hope of success (Solberg Nes et al., 2011). Regardless of dire circumstances, optimism is a healthy response to challenge (Rasmussen et al., 2009). People are more likely to try to solve problems when they are optimistic—when they believe they have some hope of succeeding. Belief in the potential success of our efforts is key to both motivation (Bandura, 2012) and creativity (Wang et al., 2018). In relation to sustainability, there is evidence that the relationship may go both ways: Where there are hopeful things happening locally, there is greater optimism (Gifford et al., 2009).

This book supports our optimism at least and our efforts to survive at best. Psychologists already apply some methods to manage creative, developmental processes in conflict-prone social ecotones. In fact, we do so widely and sometimes to good effect. There are also examples of some psychological ideas being applied to managing the growth and success of sustainable enterprises (Akemu et al., 2016; Amankwah-Amoah et al., 2019; X. Chen & Wu, 2019; Choi & Gray, 2008). So, there are reasons to assume that we will succeed.

OUR CHANCES ARE PRETTY GOOD

Can we use the same social hack to manage our social hacking better? Are we capable of using the same processes that have gotten us into trouble as ways to get us back out of trouble?

The human hack is based on creative new organizing that we now do all the time. It is the equivalent of making ourselves adaptive generalists: We can socially organize our way out of our adaptive problems because that's what social speciation does—it moderates adaptive pressures. We may also need to take a new approach to social adaptation if we want to get out of our holes, but I am convinced that we can work with the same urge to change our environment that got us where we are to be even better adapted—to sustain our adaptation for future earth inhabitants.

This book uses what we have learned so far about the psychological processes of social speciation to propose potentially fruitful ways to build our way out of our building. By taking a constructive, psychologically realistic scientist–practitioner approach, we have the means to guide social mapmaking toward a bright future. We can inject the voices of future planet mates into the social decisions of which we are part. Following Rachel Carson's (1962) call, we have all the necessary methods to "manage ourselves."

Dreaming of a better future is the essential first step. After that, it's thinking about and talking with the "right" people about how to organize to make the dream happen. We are not just reactive adapters. We are evolutionary hackers trying to figure out how to hack our hack. We're trying to build our way out of our building. We can manage the social speciation that has led to our problems to solve these same problems.

References

Abrahamse, W., Steg, L., Vlek, C., & Rothengatter, T. (2005). A review of intervention studies aimed at household energy conservation. *Journal of Environmental Psychology*, *25*(3), 273–291. https://doi.org/10.1016/j.jenvp.2005.08.002

Abwe, E. E., Morgan, B. J., Tchiengue, B., Kentatchime, F., Doudja, R., Ketchen, M. E., Teguia, E., Ambahe, R., Venditti, D. M., Mitchell, M. W., Fosso, B., Mounga, A., Fotso, R. C., & Gonder, M. K. (2019). Habitat differentiation among three Nigeria–Cameroon chimpanzee (Pan troglodytes ellioti) populations. *Ecology & Evolution*, *9*(3), 1489–1500.

Ackerman, J. (2020). *The bird way: A new look at how birds talk, work, play, parent, and think*. Penguin.

Adams, D. (2007). *The hitchhiker's guide to the galaxy*. Del Ray. (Original work published 1979)

Akemu, O., Whiteman, G., & Kennedy, S. (2016). Social enterprise emergence from social movement activism: The Fairphone case. *Journal of Management Studies*, *53*(5), 846–877. https://doi.org/10.1111/joms.12208

Alessandri, T. M., Mammen, J., & Eddleston, K. (2018). Managerial incentives, myopic loss aversion, and firm risk: A comparison of family and non-family firms. *Journal of Business Research*, *91*, 19–27. https://doi.org/10.1016/j.jbusres.2018.05.030

Amankwah-Amoah, J., Danso, A., & Adomako, S. (2019). Entrepreneurial orientation, environmental sustainability and new venture performance: Does stakeholder integration matter? *Business Strategy and the Environment*, *28*(1), 79–87. https://doi.org/10.1002/bse.2191

Amel, E. L., Manning, C. M., & Scott, B. A. (2009). Mindfulness and sustainable behavior: Pondering attention and awareness as means for increasing green behavior. *Ecopsychology*, *1*(1), 14–25. https://doi.org/10.1089/eco.2008.0005

REFERENCES

Apel, R. (2013). Sanctions, perceptions, and crime: Implications for criminal deterrence. *Journal of Quantitative Criminology, 29*(1), 67–101. https://doi.org/10.1007/s10940-012-9170-1

Archer, S., Buxton, S., & Sheffield, D. (2015). The effect of creative psychological interventions on psychological outcomes for adult cancer patients: A systematic review of randomised controlled trials. *Psycho-Oncology, 24*(1), 1–10. https://doi.org/10.1002/pon.3607

Argyris, C. (1982). The executive mind and double-loop learning. *Organizational Dynamics, 11*(2), 5–22.

Arioli, M., & Canessa, N. (2019). Neural processing of social interaction: Coordinate-based meta-analytic evidence from human neuroimaging studies. *Human Brain Mapping, 40*(13), 3712–3737. https://doi.org/10.1002/hbm.24627

Ashforth, B. E., & Humphrey, R. H. (1993). Emotional labor in service roles: The influence of identity. *Academy of Management Review, 18*(1), 88–115. https://doi.org/10.5465/amr.1993.3997508

Ask, K., & Granhag, P. A. (2007). Hot cognition in investigative judgments: The differential influence of anger and sadness. *Law and Human Behavior, 31*(6), 537–551. https://doi.org/10.1007/s10979-006-9075-3

Austin, R. L., & Fitzgerald, S. T. (2018). "I come back a better person": Identity construction and maintenance at a regional burn festival. *Sociological Inquiry, 88*(4), 599–625. https://doi.org/10.1111/soin.12226

Baas, M., Nijstad, B. A., Boot, N. C., & De Dreu, C. K. W. (2016). Mad genius revisited: Vulnerability to psychopathology, biobehavioral approach-avoidance, and creativity. *Psychological Bulletin, 142*(6), 668–692. https://doi.org/10.1037/bul0000049

Baltes, B. B., Wynne, K., Sirabian, M., Krenn, D., & De Lange, A. (2014). Future time perspective, regulatory focus, and selection, optimization, and compensation: Testing a longitudinal model. *Journal of Organizational Behavior, 35*(8), 1120–1133. https://doi.org/10.1002/job.1970

Balzer, W. K., Sulsky, L. M., Hammer, L. B., & Sumner, K. E. (1992). Task information, cognitive information, or functional validity information: Which components of cognitive feedback affect performance? *Organizational Behavior and Human Decision Processes, 53*(1), 35–54. https://doi.org/10.1016/0749-5978(92)90053-A

Bandura, A. (2012). On the functional properties of perceived self-efficacy revisited. *Journal of Management, 38*(1), 9–44. https://doi.org/10.1177/0149206311410606

Barry, J. M. (1997). *Rising tide*. Simon & Schuster.

Bartram, D. (2005). The great eight competencies: A criterion-centric approach to validation. *Journal of Applied Psychology, 90*(6), 1185–1203. https://doi.org/10.1037/0021-9010.90.6.1185

Bazarova, N. N., & Yuan, Y. C. (2013). Expertise recognition and influence in intercultural groups: Differences between face-to-face and computer-mediated communication. *Journal of Computer-Mediated Communication, 18*(4), 437–453. https://doi.org/10.1111/jcc4.12018

BBC. (2010, November 26). *Hans Rosling's 200 countries, 200 years, 4 minutes—The joy of stats—BBC 4* [Video]. YouTube. https://www.youtube.com/watch?v=jbkSRLYSojo

Beaulieu-Prévost, D., Cormier, M., Heller, S. M., Nelson-Gal, D., & McRae, K. (2019). Welcome to wonderland? A population study of intimate experiences and safe sex at a transformational mass gathering (Burning Man). *Archives of Sexual Behavior, 48*(7), 2055–2073. https://doi.org/10.1007/s10508-019-01509-9

Bell, S. T. (2007). Deep-level composition variables as predictors of team performance: A meta-analysis. *Journal of Applied Psychology, 92*(3), 595–615. https://doi.org/10.1037/0021-9010.92.3.595

Bendell, J. (2018). *Deep adaptation: A map for navigating climate tragedy.* https://mahb.stanford.edu/wp-content/uploads/2018/08/deepadaptation.pdf

Berg, C. A., Sewell, K. K., Hughes Lansing, A. E., Wilson, S. J., & Brewer, C. (2016). A developmental perspective to dyadic coping across adulthood. In J. Bookwala (Ed.), *Couple relationships in the middle and later years: Their nature, complexity, and role in health and illness* (pp. 259–280). American Psychological Association. https://doi.org/10.1037/14897-014

Bezrukova, K., Spell, C. S., Perry, J. L., & Jehn, K. A. (2016). A meta-analytical integration of over 40 years of research on diversity training evaluation. *Psychological Bulletin, 142*(11), 1227–1274. https://doi.org/10.1037/bul0000067

Bhatt, B., Qureshi, I., & Riaz, S. (2019). Social entrepreneurship in non-munificent institutional environments and implications for institutional work: Insights from China. *Journal of Business Ethics, 154*(3), 605–630. https://doi.org/10.1007/s10551-017-3451-4

Blume, B. D., Ford, J., Baldwin, T. T., & Huang, J. L. (2010). Transfer of training: A meta-analytic review. *Journal of Management, 36*(4), 1065–1105. https://doi.org/10.1177/0149206309352880

Booth, R. F., Bucky, S. F., & Berry, N. H. (1978). Predictors of psychiatric illness among Navy hospital corpsmen. *Journal of Clinical Psychology, 34*(2), 305–308. https://doi.org/10.1002/1097-4679(197804)34:2<305::AID-JCLP2270340208>3.0.CO;2-1

Bougon, M., Weick, K., & Binkhorst, D. (1977). Cognition in organizations: An analysis of the Utrecht Jazz Orchestra. *Administrative Science Quarterly, 22*(4), 606–639. https://doi.org/10.2307/2392403

Bowman, N. A. (2013). How much diversity is enough? The curvilinear relationship between college diversity interactions and first-year student outcomes.

Research in Higher Education, 54(8), 874–894. https://doi.org/10.1007/s11162-013-9300-0

Brauer, M. F., & Wiersema, M. F. (2012). Industry divestiture waves: How a firm's position influences investor returns. *Academy of Management Journal, 55*(6), 1472–1492. https://doi.org/10.5465/amj.2010.1099

Breuer, C., Hüffmeier, J., & Hertel, G. (2016). Does trust matter more in virtual teams? A meta-analysis of trust and team effectiveness considering virtuality and documentation as moderators. *Journal of Applied Psychology, 101*(8), 1151–1177. https://doi.org/10.1037/apl0000113

Bronfman, Z. Z., Ginsburg, S., & Jablonka, E. (2014). Shaping the learning curve: Epigenetic dynamics in neural plasticity. *Frontiers in Integrative Neuroscience, 8*, 55. https://doi.org/10.3389/fnint.2014.00055

Brotheridge, C. M., & Grandey, A. A. (2002). Emotional labor and burnout: Comparing two perspectives of 'people work.' *Journal of Vocational Behavior, 60*(1), 17–39. https://doi.org/10.1006/jvbe.2001.1815

Brown, S. G., Hill, N. S., & Lorinkova, N. M. (2021). Leadership and virtual team performance: A meta-analytic investigation. *European Journal of Work and Organizational Psychology, 30*(5), 672–685. https://doi.org/10.1080/1359432X.2021.1914719

Bühler, J. L., & Nikitin, J. (2020). Sociohistorical context and adult social development: New directions for 21st century research. *American Psychologist, 75*(4), 457–469. https://doi.org/10.1037/amp0000611

Burgos-Robles, A., Gothard, K. M., Monfils, M. H., Morozov, A., & Vicentic, A. (2019). Conserved features of anterior cingulate networks support observational learning across species. *Neuroscience and Biobehavioral Reviews, 107*, 215–228. https://doi.org/10.1016/j.neubiorev.2019.09.009

Burke, C. S., Stagl, K. C., Klein, C., Goodwin, G. F., Salas, E., & Halpin, S. M. (2006). What type of leadership behaviors are functional in teams? A meta-analysis. *The Leadership Quarterly, 17*(3), 288–307. https://doi.org/10.1016/j.leaqua.2006.02.007

Burnette, J. L., O'Boyle, E. H., VanEpps, E. M., Pollack, J. M., & Finkel, E. J. (2013). Mind-sets matter: A meta-analytic review of implicit theories and self-regulation. *Psychological Bulletin, 139*(3), 655–701. https://doi.org/10.1037/a0029531

Byers, J. A., & Kitchen, D. W. (1988). Mating system shift in a pronghorn population. *Behavioral Ecology and Sociobiology, 22*, 355–360.

Byron, K., Khazanchi, S., & Nazarian, D. (2010). The relationship between stressors and creativity: A meta-analysis examining competing theoretical models. *Journal of Applied Psychology, 95*(1), 201–212. https://doi.org/10.1037/a0017868

REFERENCES

Carson, R. (1962). *Silent spring*. Houghton-Mifflin.

Carvacho, H., Zick, A., Haye, A., González, R., Manzi, J., Kocik, C., & Bertl, M. (2013). On the relation between social class and prejudice: The roles of education, income, and ideological attitudes. *European Journal of Social Psychology, 43*(4), 272–285. https://doi.org/10.1002/ejsp.1961

Case, T., Holt, R., McPeek, M., & Keitt, T. (2005). The community context of species' borders: Ecological and evolutionary perspectives. *Oikos, 108*(1), 28–46. https://doi.org/10.1111/j.0030-1299.2005.13148.x

Castro, P. (2012). Legal innovation for social change: Exploring change and resistance to different types of sustainability laws. *Political Psychology, 33*(1), 105–121. https://doi.org/10.1111/j.1467-9221.2011.00863.x

Čehajić-Clancy, S., & Bilewicz, M. (2021). Moral-exemplar intervention: A new paradigm for conflict resolution and intergroup reconciliation. *Current Directions in Psychological Science, 30*(4), 335–342. https://doi.org/10.1177/09637214211013001

Chen, K. K. (2012). Artistic prosumption: Cocreative destruction at burning man. *American Behavioral Scientist, 56*(4), 570–595. https://doi.org/10.1177/0002764211429362

Chen, X., & Wu, L. (2019). Psychological capital in food safety social co-governance. *Frontiers in Psychology, 10*, 1387. https://doi.org/10.3389/fpsyg.2019.01387

Choain, L., & Malzy, T. (2019). Leading change through your creative class. *Journal of Organizational Change Management, 32*(3), 377–384. https://doi.org/10.1108/JOCM-04-2017-0118

Choi, D. Y., & Gray, E. R. (2008). The venture development process of "sustainable" entrepreneurs. *Management Research News, 31*(8), 558–569. https://doi.org/10.1108/01409170810892127

Church, A., Waclawski, J., McHenry, J. J., & McKenna, D. (1998). Organization development in high performing companies: An in-depth look at the role of O.D. in Microsoft. *Organization Development Journal, 16*(3), 16.

Cialdini, R. B. (1988). *Influence: Science and practice* (2nd ed.). Scott, Foresman.

Cialdini, R. B. (2003). Crafting normative messages to protect the environment. *Current Directions in Psychological Science, 12*(4), 105–109. https://doi.org/10.1111/1467-8721.01242

Cialdini, R. B. (2007). *Influence: The psychology of persuasion*. HarperCollins.

Clayton, S., Devine-Wright, P., Swim, J., Bonnes, M., Steg, L., Whitmarsh, L., & Carrico, A. (2016). Expanding the role for psychology in addressing environmental challenges. *American Psychologist, 71*(3), 199–215. https://doi.org/10.1037/a0039482

Colarelli, S. M., & Arvey, R. D. (2015). *The biological foundations of organizational behavior*. University of Chicago Press.

Cole, S. A. (2013). Implementing counter-measures against confirmation bias in forensic science. *Journal of Applied Research in Memory and Cognition*, 2(1), 61–62. https://doi.org/10.1016/j.jarmac.2013.01.011

Collins, S. E., Clifasefi, S. L., Stanton, J., The Leap Advisory Board, Straits, K. J. E., Gil-Kashiwabara, E., Rodriguez Espinosa, P., Nicasio, A. V., Andrasik, M. P., Hawes, S. M., Miller, K. A., Nelson, L. A., Orfaly, V. E., Duran, B. M., & Wallerstein, N. (2018). Community-based participatory research (CBPR): Towards equitable involvement of community in psychology research. *American Psychologist*, 73(7), 884–898. https://doi.org/10.1037/amp0000167

Converse, B. A., Hancock, P. I., Klotz, L. E., Clarens, A. F., & Adams, G. S. (2021). If humans design the planet: A call for psychological scientists to engage with climate engineering. *American Psychologist*, 76(5), 768–780. https://doi.org/10.1037/amp0000656

Conway, J. M., Amel, E. L., & Gerwien, D. P. (2009). Teaching and learning in the social context: A meta-analysis of service learning's effects on academic, personal, social, and citizenship outcomes. *Teaching of Psychology*, 36(4), 233–245. https://doi.org/10.1080/00986280903172969

Cooke, G. M., Landguth, E. L., & Beheregaray, L. B. (2014). Riverscape genetics identifies replicated ecological divergence across an Amazonian ecotone. *Evolution*, 68(7), 1947–1960. https://doi.org/10.1111/evo.12410

Coovert, M. D., & McNelis, K. (1992). Team decision making and performance: A review and proposed modeling approach employing Petri nets. In R. W. Swezey & E. Salas (Eds.), *Teams: Their training and performance* (pp. 247–280). Ablex.

Cornelissen, J. P., Akemu, O., Jonkman, J. G. F., & Werner, M. D. (2021). Building character: The formation of a hybrid organizational identity in a social enterprise. *Journal of Management Studies*, 58(5), 1294–1330. https://doi.org/10.1111/joms.12640

Costa, P. T., Jr., & McCrae, R. R. (1988). Personality in adulthood: A six-year longitudinal study of self-reports and spouse ratings on the NEO Personality Inventory. *Journal of Personality and Social Psychology*, 54(5), 853–863. https://doi.org/10.1037/0022-3514.54.5.853

Costa, P. T., Jr., & McCrae, R. R. (1989). *The NEO-PI/NEO-FFI manual supplement*. Psychological Assessment Resources.

Cozzarelli, C., Wilkinson, A. V., & Tagler, M. J. (2001). Attitudes toward the poor and attributions for poverty. *Journal of Social Issues*, 57(2), 207–227. https://doi.org/10.1111/0022-4537.00209

Cracco, E., Bardi, L., Desmet, C., Genschow, O., Rigoni, D., De Coster, L., Radkova, I., Deschrijver, E., & Brass, M. (2018). Automatic imitation: A meta-analysis. *Psychological Bulletin*, 144(5), 453–500. https://doi.org/10.1037/bul0000143

Cracco, E., Genschow, O., Radkova, I., & Brass, M. (2018). Automatic imitation of pro- and antisocial gestures: Is implicit social behavior censored? *Cognition, 170,* 179–189. https://doi.org/10.1016/j.cognition.2017.09.019

Crain, W. (2000). *Theories of development: Concepts and applications* (4th ed.). Prentice-Hall.

Crain, W. (2010). *Theories of development: Concepts and applications* (5th ed.). Psychology Press.

Crook, T. R., Combs, J. G., Ketchen, D. J., Jr., & Aguinis, H. (2013). Organizing around transaction costs: What have we learned and where do we go from here? *The Academy of Management Perspectives, 27*(1), 63–79. https://doi.org/10.5465/amp.2012.0008

Cruwys, T., Dingle, G. A., Hornsey, M. J., Jetten, J., Oei, T. P. S., & Walter, Z. C. (2014). Social isolation schema responds to positive social experiences: Longitudinal evidence from vulnerable populations. *British Journal of Clinical Psychology, 53*(3), 265–280. https://doi.org/10.1111/bjc.12042

Cuevas, J. A., & Dawson, B. L. (2021). An integrated review of recent research on the relationships between religious belief, political ideology, authoritarianism, and prejudice. *Psychological Reports, 124*(3), 977–1014. https://doi.org/10.1177/0033294120925392

da Costa, S., Páez, D., Sánchez, F., Garaigordobil, M., & Gondim, S. (2015). Personal factors of creativity: A second order meta-analysis. *Journal of Work and Organizational Psychology, 31*(3), 165–173. https://doi.org/10.1016/j.rpto.2015.06.002

Davies, M. (2011). Introduction to the special issue on critical thinking in higher education. *Higher Education Research & Development, 30*(3), 255–260. https://doi.org/10.1080/07294360.2011.562145

de Wit, F. R. C., Greer, L. L., & Jehn, K. A. (2012). The paradox of intragroup conflict: A meta-analysis. *Journal of Applied Psychology, 97*(2), 360–390. https://doi.org/10.1037/a0024844

DeChurch, L. A., Mesmer-Magnus, J. R., & Doty, D. (2013). Moving beyond relationship and task conflict: Toward a process-state perspective. *Journal of Applied Psychology, 98*(4), 559–578. https://doi.org/10.1037/a0032896

Desivilya, H. S., Somech, A., & Lidgoster, H. (2010). Innovation and conflict management in work teams: The effects of team identification and task and relationship conflict. *Negotiation and Conflict Management Research, 3*(1), 28–48. https://doi.org/10.1111/j.1750-4716.2009.00048.x

Diamond, J. (1997). *Guns, germs, and steel.* Norton.

Dionne, G. (2013). Risk management: History, definition, and critique. *Risk Management & Insurance Review, 16*(2), 147–166. https://doi.org/10.1111/rmir.12016

REFERENCES

Dokko, G., Kane, A. A., & Tortoriello, M. (2014). One of us or one of my friends: How social identity and tie strength shape the creative generativity of boundary-spanning ties. *Organization Studies, 35*(5), 703–726. https://doi.org/10.1177/0170840613508397

Domínguez, D. G., García, D., Martínez, D. A., & Hernandez-Arriaga, B. (2020). Leveraging the power of mutual aid, coalitions, leadership, and advocacy during COVID-19. *American Psychologist, 75*(7), 909–918. https://doi.org/10.1037/amp0000693

Donahue, M. J. (1985). Intrinsic and extrinsic religiousness: Review and meta-analysis. *Journal of Personality and Social Psychology, 48*(2), 400–419. https://doi.org/10.1037/0022-3514.48.2.400

Dong, B., Xu, H., Luo, J., Nicol, C. D., & Liu, W. (2020). Many roads lead to Rome: How entrepreneurial orientation and trust boost the positive network range and entrepreneurial performance relationship. *Industrial Marketing Management, 88*, 173–185. https://doi.org/10.1016/j.indmarman.2020.04.016

Dono, J., Webb, J., & Richardson, B. (2010). The relationship between environmental activism, pro-environmental behaviour and social identity. *Journal of Environmental Psychology, 30*(2), 178–186. https://doi.org/10.1016/j.jenvp.2009.11.006

Drescher, J. (2003). Gold or lead? Introductory remarks on conversions. *Journal of Gay & Lesbian Psychotherapy, 7*(3), 1–13. https://doi.org/10.1300/J236v07n03_01

Dunford, R. (1987). The suppression of technology as a strategy for controlling resource dependence. *Administrative Science Quarterly, 32*(4), 512–525. https://doi.org/10.2307/2392881

Ekman, P. (1992). An argument for basic emotions. *Cognition and Emotion, 6*(3–4), 169–200. https://doi.org/10.1080/02699939208411068

Elliot, A. J., Eder, A. B., & Harmon-Jones, E. (2013). Approach–avoidance motivation and emotion: Convergence and divergence. *Emotion Review, 5*(3), 308–311. https://doi.org/10.1177/1754073913477517

Feng, C., Eickhoff, S. B., Li, T., Wang, L., Becker, B., Camilleri, J. A., Hétu, S., & Luo, Y. (2021). Common brain networks underlying human social interactions: Evidence from large-scale neuroimaging meta-analysis. *Neuroscience and Biobehavioral Reviews, 126*, 289–303. https://doi.org/10.1016/j.neubiorev.2021.03.025

Fernando, J. W., Burden, N., Ferguson, A., O'Brien, L. V., Judge, M., & Kashima, Y. (2018). Functions of utopia: How utopian thinking motivates societal engagement. *Personality and Social Psychology Bulletin, 44*(5), 779–792. https://doi.org/10.1177/0146167217748604

Festinger, L., Riecken, H. W., & Schachter, S. (1964). *When prophecy fails*. Harper Torchbooks.

Fisher, A. (2018, July 10). "Google was not a normal place": Brin, Page, and Mayer on the accidental birth of the company that changed everything. *Vanity Fair*. https://www.vanityfair.com/news/2018/07/valley-of-genius-excerpt-google

Fisher, G., Kotha, S., & Lahiri, A. (2016). Changing with the times: An integrated view of identity, legitimacy, and new venture life cycles. *Academy of Management Review, 41*(3), 383–409. https://doi.org/10.5465/amr.2013.0496

Fisher, M., Aikens, J., Kennell, R., Ayala, N., Klimoski, R. J., & Jones, R. G. (2019, April 4–6). *Selection feedback derived from biodata correlates of long-term strategic perspective* [Poster presentation]. Society for Industrial and Organizational Psychology 34th Annual Conference, Washington, DC/National Harbor, MD, United States.

Florida, R. (2005). *Cities and the creative class*. Routledge. https://doi.org/10.4324/9780203997673

Foitzik, S., & Fritsche, O. (2019). *Empire of ants*. The Experiment.

Folke, C., Colding, J., & Berkes, F. (2003). Synthesis: Building resilience and adaptive capacity in socio-ecological systems. In F. Berkes, C. Folke, & J. Colding (Eds.), *Navigating social–ecological systems: Building resilience for complexity and change* (pp. 352–387). Cambridge University Press.

Ford, J. K. (2021). *Learning in organizations*. Routledge.

Frankopan, P. (2017). *The silk roads*. Bloomsbury.

French, K. A., Dumani, S., Allen, T. D., & Shockley, K. M. (2018). A meta-analysis of work-family conflict and social support. *Psychological Bulletin, 144*(3), 284–314. https://doi.org/10.1037/bul0000120

French, K. A., & Shockley, K. M. (2020). Formal and informal supports for managing work and family. *Current Directions in Psychological Science, 29*(2), 207–216. https://doi.org/10.1177/0963721420906218

Fritsche, I., Cohrs, J. C., Kessler, T., & Bauer, J. (2012). Global warming is breeding social conflict: The subtle impact of climate change threat on authoritarian tendencies. *Journal of Environmental Psychology, 32*(1), 1–10. https://doi.org/10.1016/j.jenvp.2011.10.002

Funder, D. C. (1989). Accuracy in personality judgment and the dancing bear. In D. Buss & N. Cantor (Eds.), *Personality research for the 1990s* (pp. 210–223). Springer-Verlag. https://doi.org/10.1007/978-1-4684-0634-4_16

Gabenesch, H., & Hunt, L. L. (1971). The relative accuracy of interpersonal perception of high and low authoritarians. *Journal of Experimental Research in Personality, 5*, 43–48.

Gaither, S. E., Apfelbaum, E. P., Birnbaum, H. J., Babbitt, L. G., & Sommers, S. R. (2018). Mere membership in racially diverse groups reduces conformity. *Social Psychological & Personality Science, 9*(4), 402–410. https://doi.org/10.1177/1948550617708013

Gallagher, V. C., Porter, T. H., & Gallagher, K. P. (2020). Sustainability change agents: Leveraging political skill and reputation. *Journal of Organizational Change Management, 33*(1), 181–195. https://doi.org/10.1108/JOCM-01-2018-0031

Gardikiotis, A. (2011). Minority influence. *Social and Personality Psychology Compass, 5*(9), 679–693. https://doi.org/10.1111/j.1751-9004.2011.00377.x

Genschow, O., van Den Bossche, S., Cracco, E., Bardi, L., Rigoni, D., & Brass, M. (2017). Mimicry and automatic imitation are not correlated. *PLOS ONE, 12*(9), Article e0183784. https://doi.org/10.1371/journal.pone.0183784

Gerlai, R. (2001). Gene targeting: Technical confounds and potential solutions in behavioral brain research. *Behavioural Brain Research, 125*(1–2), 13–21. https://doi.org/10.1016/S0166-4328(01)00282-0

Ghosh, R. (2014). Antecedents of mentoring support: A meta-analysis of individual, relational, and structural or organizational factors. *Journal of Vocational Behavior, 84*(3), 367–384. https://doi.org/10.1016/j.jvb.2014.02.009

Gifford, R. (2011). The dragons of inaction: Psychological barriers that limit climate change mitigation and adaptation. *American Psychologist, 66*(4), 290–302. https://doi.org/10.1037/a0023566

Gifford, R., Scannell, L., Kormos, C., Smolova, L., Biel, A., Boncu, S., Corral, V., Güntherf, H., Hanyu, K., Hine, D., Kaiser, F. G., Korpela, K., Lima, L. M., Mertig, A. G., Mira, R. G., Moser, G., Pinheiro, J. Q., Saini, S., Sako, T., . . . Uzzell, D. (2009). Temporal pessimism and spatial optimism in environmental assessments: An 18-nation study. *Journal of Environmental Psychology, 29*(1), 1–12. https://doi.org/10.1016/j.jenvp.2008.06.001

Gittleman, J. L., Gardner, P. C., Haile, E., Sampson, J. M., Cigularov, K. P., Ermann, E. D., Stafford, P., & Chen, P. Y. (2010). CityCenter and Cosmopolitan Construction Projects, Las Vegas, Nevada: Lessons learned from the use of multiple sources and mixed methods in a safety needs assessment. *Journal of Safety Research, 41*(3), 263–281. https://doi.org/10.1016/j.jsr.2010.04.004

Gladstein, D. L., & Reilly, N. P. (1985). Group decision making under threat: The tycoon game. *Academy of Management Journal, 28*(3), 613–627.

Gluck, M. A., & Myers, C. E. (2001). *Gateway to memory: An introduction to neural network modeling of the hippocampus and learning.* MIT Press.

Gonzalez-Mulé, E., Kim, M. M., & Ryu, J. W. (2021). A meta-analytic test of multiplicative and additive models of job demands, resources, and stress. *Journal of Applied Psychology, 106*(9), 1391–1411. https://doi.org/10.1037/apl0000840

Gowlett, J. A. (2016). The discovery of fire by humans: A long and convoluted process. *Philosophical Transactions of the Royal Society B: Biological Sciences, 371*(1696). https://doi.org/10.1098/rstb.2015.0164

Gregersen, T., Doran, R., Böhm, G., Tvinnereim, E., & Poortinga, W. (2020). Political orientation moderates the relationship between climate change beliefs and worry about climate change. *Frontiers in Psychology, 11*, 1573. https://doi.org/10.3389/fpsyg.2020.01573

Grutterink, H., Van der Vegt, G. S., Molleman, E., & Jehn, K. A. (2013). Reciprocal expertise affirmation and shared expertise perceptions in work teams: Their implications for coordinated action and team performance. *Applied Psychology, 62*(3), 359–381. https://doi.org/10.1111/j.1464-0597.2012.00484.x

Guan, L.-L., Xiang, H., Wu, X.-M., Ma, N., Wu, B.-M., Cheng, W.-H., Ma, H., & Xiao, Z.-P. (2011). Needs assessment for the psychosocial support at the severely-affected counties after Wenchuan earthquake. *Chinese Mental Health Journal, 25*(2), 107–112.

Guassi Moreira, J. F., Tashjian, S. M., Galván, A., & Silvers, J. A. (2021). Computational and motivational mechanisms of human social decision making involving close others. *Journal of Experimental Social Psychology, 93*, Article 104086. https://doi.org/10.1016/j.jesp.2020.104086

Guenter, H., Gardner, W. L., Davis McCauley, K., Randolph-Seng, B., & Prabhu, V. P. (2017). Shared authentic leadership in research teams: Testing a multiple mediation model. *Small Group Research, 48*(6), 719–765. https://doi.org/10.1177/1046496417732403

Gulledge, A. T., & Kawaguchi, Y. (2007). Phasic cholinergic signaling in the hippocampus: Functional homology with the neocortex? *Hippocampus, 17*(5), 327–332. https://doi.org/10.1002/hipo.20279

Halford, G. S., & McCredden, J. E. (1998). Cognitive science questions for cognitive development: The concepts of learning, analogy, and capacity. *Learning and Instruction, 8*(4), 289–308. https://doi.org/10.1016/S0959-4752(97)00023-6

Hall, D. L., Matz, D. C., & Wood, W. (2010). Why don't we practice what we preach? A meta-analytic review of religious racism. *Personality and Social Psychology Review, 14*(1), 126–139. https://doi.org/10.1177/1088868309352179

Hamm, M., & Drossel, B. (2017). Habitat heterogeneity hypothesis and edge effects in model metacommunities. *Journal of Theoretical Biology, 426*, 40–48. https://doi.org/10.1016/j.jtbi.2017.05.022

Harari, N. (2014). *Sapiens: A brief history of humankind*. Harvill Secker.

Harrington, J. (1656). *The Republic of Oceana*. Pakeman.

Harris, P. L. (2019). Affective social learning: From nature to culture. In D. Dukes & F. Clément (Eds.), *Foundations of affective social learning: Conceptualizing the social transmission of value* (pp. 69–86). Cambridge University Press. https://doi.org/10.1017/9781108661362.004

Heaven, P. C. L., Ciarrochi, J., & Leeson, P. (2011). Cognitive ability, right-wing authoritarianism, and social dominance orientation: A five-year longitudinal study amongst adolescents. *Intelligence, 39*(1), 15–21. https://doi.org/10.1016/j.intell.2010.12.001

Hedeen, T., & Kelly, R. (2009). Challenging conventions in challenging conditions: Thirty-minute mediations at Burning Man. *Conflict Resolution Quarterly, 27*(1), 107–119. https://doi.org/10.1002/crq.250

Hedge, C., Powell, G., Bompas, A., Vivian-Griffiths, S., & Sumner, P. (2018). Low and variable correlation between reaction time costs and accuracy costs explained by accumulation models: Meta-analysis and simulations. *Psychological Bulletin, 144*(11), 1200–1227. https://doi.org/10.1037/bul0000164

Hedge, J. W., & Pulakos, E. D. (2002). Grappling with implementation: Some preliminary thoughts and relevant research. In J. W. Hedge & E. D. Pulakos (Eds.), *Implementing organizational interventions: Steps, processes, and best practices* (pp. 1–11). Jossey-Bass.

Heers, A. M. (2016). New perspectives on the ontogeny and evolution of avian locomotion. *Integrative and Comparative Biology, 56*(3), 428–441. https://doi.org/10.1093/icb/icw065

Henrich, J. (2015). *The secret of our success*. Princeton University Press. https://doi.org/10.2307/j.ctvc77f0d

Henrich, J., & Muthukrishna, M. (2021). The origins and psychology of human cooperation. *Annual Review of Psychology, 72*(1), 207–240. https://doi.org/10.1146/annurev-psych-081920-042106

Hernández, B., Martín, A. M., Ruiz, C., & Hidalgo, M. (2010). The role of place identity and place attachment in breaking environmental protection laws. *Journal of Environmental Psychology, 30*(3), 281–288. https://doi.org/10.1016/j.jenvp.2010.01.009

Herzenstein, M., & Posavac, S. S. (2019). When charity begins at home: How personal financial scarcity drives preference for donating locally at the expense of global concerns. *Journal of Economic Psychology, 73*, 123–135. https://doi.org/10.1016/j.joep.2019.06.002

Hilscher, S. (2016). Krieg, Kultur und Psyche: Ein subjektiver Erlebnisbericht [War, culture and mind]. *Nervenheilkunde: Zeitschrift für Interdisziplinaere Fortbildung, 35*(6), 375–377. https://doi.org/10.1055/s-0037-1616396

Hoover, J. D. (2008). Realizing the artful in management education and development: Smoldering examples from the Burning Man Project. *Journal of Management & Organization, 14*(5), 535–547. https://doi.org/10.5172/jmo.837.14.5.535

Hornsey, M. J., & Fielding, K. S. (2017). Attitude roots and Jiu Jitsu persuasion: Understanding and overcoming the motivated rejection of science. *American Psychologist, 72*(5), 459–473. https://doi.org/10.1037/a0040437

Hothersall, D. (1983). *History of psychology*. Random House.

Hothersall, D. (1995). *History of psychology* (3rd ed.). McGraw-Hill.

Huang, J. L., Blume, B. D., Ford, J. K., & Baldwin, T. T. (2015). A tale of two transfers: Disentangling maximum and typical transfer and their respective predictors. *Journal of Business and Psychology, 30*(4), 709–732. https://doi.org/10.1007/s10869-014-9394-1

Hughes, A. M., Zajac, S., Woods, A. L., & Salas, E. (2020). The role of work environment in training sustainment: A meta-analysis. *Human Factors, 62*(1), 166–183. https://doi.org/10.1177/0018720819845988

Iyengar, S. (2013). Comparative and evolutionary aspects of cognition. In P. N. Tandon, R. C. Tripathi, & N. Srinivasan (Eds.), *Expanding horizons of the mind sciences* (pp. 249–265). Novinka/Nova Science Publishers.

Jablonski, N., & Chaplin, G. (1992). The origin of hominid bipedalism re-examined. *Archaeology in Oceania, 27*(3), 113–119. https://doi.org/10.1002/j.1834-4453.1992.tb00294.x

Jackendoff, R., & Wittenberg, E. (2017). Linear grammar as a possible stepping-stone in the evolution of language. *Psychonomic Bulletin & Review, 24*(1), 219–224. https://doi.org/10.3758/s13423-016-1073-y

Jiang, J. Y., Zhang, X., & Tjosvold, D. (2013). Emotion regulation as a boundary condition of the relationship between team conflict and performance: A multilevel examination. *Journal of Organizational Behavior, 34*(5), 714–734. https://doi.org/10.1002/job.1834

Jones, R. G. (2015). *Psychology of sustainability: An applied perspective*. Taylor & Francis/Routledge.

Jones, R. G. (2017). *An applied approach to the psychology of sustainability*. Oxford University Press. https://oxfordre.com/psychology/view/10.1093/acrefore/9780190236557.001.0001/acrefore-9780190236557-e-76

Jones, R. G. (2018). *Technical report for the Strategic Preferences Inventory* [Unpublished manuscript]. Psychology Department, Missouri State University.

Jones, R. G. (2020). *Applied psychology of sustainability*. Taylor & Francis/Routledge. https://doi.org/10.4324/9780429488382

Jones, R. G., Corwin, E., Anderson, S., & McKenna, M. (2016, April 14–16). *Long term strategic thinking: Predictor construct for performance and sustainability* [Poster presentation]. Society for Industrial and Organizational Psychology, 31st Annual Conference, Anaheim, CA, United States.

Jones, R. G., Levesque, C., & Masuda, A. (2003). Emotional displays and social identity: Emotional investment in organizations. In G. Skarlicki & D. Steiner (Eds.), *Social values in organizations* (pp. 205–220). Information Age.

Jones, R. G., & Lindley, W. D. (1998). Issues in the transition to teams. *Journal of Business and Psychology, 13*, 31–40. https://doi.org/10.1023/A:1022966915545

Jones, R. G., & Parameswaran, G. (2005). Predicting the human weather: How differentiation and contextual complexity affect behavior prediction. In K. Richardson

(Ed.), *Managing the complex: Philosophy, theory, and applications* (pp. 183–200). Information Age.

Jones, R. G., Stevens, M. J., & Fischer, D. L. (2000). Selection in team contexts. In J. F. Kehoe (Ed.), *Managing selection in changing organizations* (pp. 210–241). Jossey-Bass.

Joshi, A. (2014). By whom and when is women's expertise recognized? The interactive effects of gender and education in science and engineering teams. *Administrative Science Quarterly, 59*(2), 202–239. https://doi.org/10.1177/0001839214528331

Jost, J. T., Glaser, J., Kruglanski, A. W., & Sulloway, F. J. (2003). Political conservatism as motivated social cognition. *Psychological Bulletin, 129*(3), 339–375. https://doi.org/10.1037/0033-2909.129.3.339

Judge, T. A., Higgins, C. A., Thoresen, C. J., & Barrick, M. A. (1999). The Big Five personality traits, general mental ability, and career success across the lifespan. *Personnel Psychology, 52*(3), 621–652. https://doi.org/10.1111/j.1744-6570.1999.tb00174.x

Kahneman, D. (2011). *Thinking, fast and slow*. Farrar, Straus & Giroux.

Katz, D., & Kahn, R. L. (1978). *The social psychology of organizations*. Wiley.

Kazantzis, N., Luong, H. K., Usatoff, A. S., Impala, T., Yew, R. Y., & Hofmann, S. G. (2018). The processes of cognitive behavioral therapy: A review of meta-analyses. *Cognitive Therapy and Research, 42*(4), 349–357. https://doi.org/10.1007/s10608-018-9920-y

Kessler, T., & Cohrs, J. C. (2008). The evolution of authoritarian processes: Fostering cooperation in large-scale groups. *Group Dynamics, 12*(1), 73–84. https://doi.org/10.1037/1089-2699.12.1.73

Kim, B., Jee, S., Lee, J., An, S., & Lee, S. M. (2018). Relationships between social support and student burnout: A meta-analytic approach. *Stress and Health, 34*(1), 127–134. https://doi.org/10.1002/smi.2771

Kirkman, B. L., Jones, R. G., & Shapiro, D. L. (2000). Why do employees resist teams? Examining the resistance barrier to work team effectiveness. *International Journal of Conflict Management, 11*(1), 74–92. https://doi.org/10.1108/eb022836

Kirkman, B. L., Shapiro, D. L., Lu, S., & McGurrin, D. P. (2016). Culture and teams. *Current Opinion in Psychology, 8*, 137–142. https://doi.org/10.1016/j.copsyc.2015.12.001

Klasmeier, K. N., & Rowold, J. (2020). A multilevel investigation of predictors and outcomes of shared leadership. *Journal of Organizational Behavior, 41*(9), 915–930. https://doi.org/10.1002/job.2477

Klein, S. R., & Huffman, A. H. (2013). I-O psychology and environmental sustainability in organizations: A natural partnership. In S. R. Klein & A. H. Huffman

(Eds.), *Green organizations: Driving change with I-O psychology* (pp. 3–16). Routledge.

Klimoski, R. J., & Jones, R. G. (1995). Staffing for effective group decision making: Key issues in matching people and teams. In R. Guzzo & E. Salas (Eds.), *Team effectiveness and decision making in organizations* (pp. 291–332). Jossey-Bass.

Klimoski, R. J., & Karol, B. L. (1976). The impact of trust on creative problem solving groups. *Journal of Applied Psychology, 61*(5), 630–633. https://doi.org/10.1037/0021-9010.61.5.630

Klimoski, R. J., & Mohammed, S. (1994). Team mental model: Construct or metaphor? *Journal of Management, 20*(2), 403–437. https://doi.org/10.1177/014920639402000206

Kluger, A. N., & DeNisi, A. (1996). The effects of feedback interventions on performance: A historical review, a meta-analysis, and a preliminary feedback intervention theory. *Psychological Bulletin, 119*(2), 254–284. https://doi.org/10.1037/0033-2909.119.2.254

Koh, B., & Leung, A. K.-y. (2019). A time for creativity: How future-oriented schemas facilitate creativity. *Journal of Experimental Social Psychology, 84,* Article 103816. https://doi.org/10.1016/j.jesp.2019.103816

Koh, G. C. H., Khoo, H. E., Wong, M. L., & Koh, D. (2008). The effects of problem-based learning during medical school on physician competency: A systematic review. *Canadian Medical Association Journal, 178*(1), 34–41. https://doi.org/10.1503/cmaj.070565

Kossek, E. E., Huang, J. L., Piszczek, M. M., Fleenor, J. W., & Ruderman, M. (2017). Rating expatriate leader effectiveness in multisource feedback systems: Cultural distance and hierarchical effects. *Human Resource Management, 56*(1), 151–172. https://doi.org/10.1002/hrm.21763

Krupat, E., Pololi, L., Schnell, E. R., & Kern, D. E. (2013). Changing the culture of academic medicine: The C-Change learning action network and its impact at participating medical schools. *Academic Medicine, 88*(9), 1252–1258. https://doi.org/10.1097/ACM.0b013e31829e84e0

Kuhn, D., Cheney, R., & Weinstock, M. (2000). The development of epistemological understanding. *Cognitive Development, 15*(3), 309–328. https://doi.org/10.1016/S0885-2014(00)00030-7

LaFrance, A. (2020, December 15). Facebook is a doomsday machine. *The Atlantic.* https://www.theatlantic.com/technology/archive/2020/12/facebook-doomsday-machine/617384/

Lambert, H. K., & McLaughlin, K. A. (2019). Impaired hippocampus-dependent associative learning as a mechanism underlying PTSD: A meta-analysis. *Neuroscience and Biobehavioral Reviews, 107,* 729–749. https://doi.org/10.1016/j.neubiorev.2019.09.024

REFERENCES

Laredo, S. A., Steinman, M. Q., Robles, C. F., Ferrer, E., Ragen, B. J., & Trainor, B. C. (2015). Effects of defeat stress on behavioral flexibility in males and females: Modulation by the mu-opioid receptor. *The European Journal of Neuroscience, 41*(4), 434–441. https://doi.org/10.1111/ejn.12824

Lautenschlager, S., Witmer, L. M., Altangerel, P., & Rayfield, E. J. (2013). Edentulism, beaks, and biomechanical innovations in the evolution of theropod dinosaurs. *Proceedings of the National Academy of Sciences, 110*(51), 20657–20662. https://doi.org/10.1073/pnas.1310711110

Lee, A., Willis, S., & Tian, A. W. (2018). Empowering leadership: A meta-analytic examination of incremental contribution, mediation, and moderation. *Journal of Organizational Behavior, 39*(3), 306–325. https://doi.org/10.1002/job.2220

Lee, M., Adbi, A., & Singh, J. (2020). Categorical cognition and outcome efficiency in impact investing decisions. *Strategic Management Journal, 41*(1), 86–107. https://doi.org/10.1002/smj.3096

Leidner, B., Tropp, L. R., & Lickel, B. (2013). Bringing science to bear—On peace, not war. *American Psychologist, 68*(7), 514–526. https://doi.org/10.1037/a0032846

Leighton, L. J., Ke, K., Zajaczkowski, E. L., Edmunds, J., Spitale, R. C., & Bredy, T. W. (2018). Experience-dependent neural plasticity, learning, and memory in the era of epitranscriptomics. *Genes, Brain & Behavior, 17*(3), Article e12426. https://doi.org/10.1111/gbb.12426

Leslie, L. M., Bono, J. E., Kim, Y. S., & Beaver, G. R. (2020). On melting pots and salad bowls: A meta-analysis of the effects of identity-blind and identity-conscious diversity ideologies. *Journal of Applied Psychology, 105*(5), 453–471. https://doi.org/10.1037/apl0000446

Lewis, P., & Wong, J. C. (2018, March 18). Facebook employs psychologist whose firm sold data to Cambridge Analytica. *The Guardian*. https://www.theguardian.com/news/2018/mar/18/facebook-cambridge-analytica-joseph-chancellor-gsr

Li, M., & Hsu, C. H. C. (2018). Customer participation in services and employee innovative behavior: The mediating role of interpersonal trust. *International Journal of Contemporary Hospitality Management, 30*(4), 2112–2131. https://doi.org/10.1108/IJCHM-08-2016-0465

Liang, S., Ye, D., & Liu, Y. (2021). The effect of perceived scarcity: Experiencing scarcity increases risk taking. *The Journal of Psychology, 155*(1), 59–89. https://doi.org/10.1080/00223980.2020.1822770

Lines, R. L. J., Pietsch, S., Crane, M., Ntoumanis, N., Temby, P., Graham, S., & Gucciardi, D. F. (2021). The effectiveness of team reflexivity interventions: A systematic review and meta-analysis of randomized controlled trials. *Sport, Exercise, and Performance Psychology, 10*(3), 438–473. https://doi.org/10.1037/spy0000251

REFERENCES

Liu, D., Jiang, K., Shalley, C. E., Keem, S., & Zhou, J. (2016). Motivational mechanisms of employee creativity: A meta-analytic examination and theoretical extension of the creativity literature. *Organizational Behavior and Human Decision Processes, 137*, 236–263. https://doi.org/10.1016/j.obhdp.2016.08.001

Liu, S., Huang, J. L., & Wang, M. (2014). Effectiveness of job search interventions: A meta-analytic review. *Psychological Bulletin, 140*(4), 1009–1041. https://doi.org/10.1037/a0035923

Lourenço, O. (2012). Piaget and Vygotsky: Many resemblances, and a crucial difference. *New Ideas in Psychology, 30*(3), 281–295. https://doi.org/10.1016/j.newideapsych.2011.12.006

Lowman, R. L. (2013). Is sustainability and ethical responsibility of I-O and consulting psychologists? In A. H. Huffman & S. R. Klein (Eds.), *Green organizations: Driving change with I-O psychology* (pp. 34–54). Routledge.

Luu, T. T., & Djurkovic, N. (2019). Paternalistic leadership and idiosyncratic deals in a healthcare context. *Management Decision, 57*(3), 621–648. https://doi.org/10.1108/MD-06-2017-0595

Lynch, M. J., Barrett, K. L., Stretesky, P. B., & Long, M. A. (2016). The weak probability of punishment for environmental offenses and deterrence of environmental offenders: A discussion based on USEPA criminal cases, 1983–2013. *Deviant Behavior, 37*(10), 1095–1109. https://doi.org/10.1080/01639625.2016.1161455

Mannen, D., Hinton, S., Kuijper, T., & Porter, T. (2012). Sustainable organizing: A multiparadigm perspective of organizational development and permaculture gardening. *Journal of Leadership & Organizational Studies, 19*(3), 355–368. https://doi.org/10.1177/1548051812442967

Marino, L. (2002). Convergence of complex cognitive abilities in cetaceans and primates. *Brain, Behavior and Evolution, 59*(1–2), 21–32. https://doi.org/10.1159/000063731

Marko, M., & Riečanský, I. (2018). Sympathetic arousal, but not disturbed executive functioning, mediates the impairment of cognitive flexibility under stress. *Cognition, 174*, 94–102. https://doi.org/10.1016/j.cognition.2018.02.004

Markus, H. R., & Kitayama, S. (2010). Cultures and selves: A cycle of mutual constitution. *Perspectives on Psychological Science, 5*(4), 420–430. https://doi.org/10.1177/1745691610375557

Martin, R., Martin, P. Y., Smith, J. R., & Hewstone, M. (2007). Majority versus minority influence and prediction of behavioral intentions and behavior. *Journal of Experimental Social Psychology, 43*(5), 763–771. https://doi.org/10.1016/j.jesp.2006.06.006

Masuda, A., & Visio, M. (2012). Nepotism practices and the work-family interface. In R. G. Jones (Ed.), *Nepotism in organizations* (pp. 147–170). Routledge.

Mathieu, M., Eschleman, K. J., & Cheng, D. (2019). Meta-analytic and multiwave comparison of emotional support and instrumental support in the workplace. *Journal of Occupational Health Psychology, 24*(3), 387–409. https://doi.org/10.1037/ocp0000135

McKenzie-Mohr, D. (2000). Fostering sustainable behavior through community-based social marketing. *American Psychologist, 55*(5), 531–537. https://doi.org/10.1037/0003-066X.55.5.531

Mesmer-Magnus, J., Niler, A. A., Plummer, G., Larson, L. E., & DeChurch, L. A. (2017). The cognitive underpinnings of effective teamwork: A continuation. *Career Development International, 22*(5), 507–519. https://doi.org/10.1108/CDI-08-2017-0140

Metzger, J. A. (2014). Adaptive defense mechanisms: Function and transcendence. *Journal of Clinical Psychology, 70*(5), 478–488. https://doi.org/10.1002/jclp.22091

Mitzinneck, B. C., & Besharov, M. L. (2019). Managing value tensions in collective social entrepreneurship: The role of temporal, structural, and collaborative compromise. *Journal of Business Ethics, 159*(2), 381–400. https://doi.org/10.1007/s10551-018-4048-2

Mohammed, S., & Dumville, B. C. (2001). Team mental models in a team knowledge framework: Expanding theory and measurement across disciplinary boundaries. *Journal of Organizational Behavior, 22*(2), 89–106. https://doi.org/10.1002/job.86

Molnar, T. S. (1967). *Utopia: The perennial heresy*. Sheed and Ward.

Montani, F., Setti, I., Sommovigo, V., Courcy, F., & Giorgi, G. (2020). Who responds creatively to role conflict? Evidence for a curvilinear relationship mediated by cognitive adjustment at work and moderated by mindfulness. *Journal of Business and Psychology, 35*(5), 621–641. https://doi.org/10.1007/s10869-019-09644-9

Moore, C., Mayer, D. M., Chiang, F. F. T., Crossley, C., Karlesky, M. J., & Birtch, T. A. (2019). Leaders matter morally: The role of ethical leadership in shaping employee moral cognition and misconduct. *Journal of Applied Psychology, 104*(1), 123–145. https://doi.org/10.1037/apl0000341

Morris, E. (1979). *The rise of Theodore Roosevelt*. Random House.

Morris, I. (2010). *Why the West rules—For now*. Farrar, Straus & Giroux.

Muchinsky, P. M. (2004). When the psychometrics of test development meet organizational realities: A conceptual framework for organizational change, examples, and recommendations. *Personnel Psychology, 57*(1), 175–209. https://doi.org/10.1111/j.1744-6570.2004.tb02488.x

Mullins, J. W., & Cummings, L. L. (1999). Situational strength: A framework for understanding the role of individuals in initiating proactive strategic change.

Journal of Organizational Change Management, 12(6), 462–479. https://doi.org/10.1108/09534819910300846

National Institute of Mental Health. (2019). *Mental illness.* https://www.nimh.nih.gov/health/statistics/mental-illness

Nicholson, N. (2015). Primal business: Evolution, kinship, and the family firm. In S. Colarelli & R. Arvey (Eds.), *Biological foundations of organizational behavior* (pp. 237–267). University of Chicago Press.

Nielsen, K. S., Clayton, S., Stern, P. C., Dietz, T., Capstick, S., & Whitmarsh, L. (2021). How psychology can help limit climate change. *American Psychologist, 76*(1), 130–144. https://doi.org/10.1037/amp0000624

Nohe, C., Meier, L. L., Sonntag, K., & Michel, A. (2015). The chicken or the egg? A meta-analysis of panel studies of the relationship between work-family conflict and strain. *Journal of Applied Psychology, 100*(2), 522–536. https://doi.org/10.1037/a0038012

No spectators: The art of Burning Man. (2018). https://americanart.si.edu/exhibitions/burning-man

Nweke, O. C., Payne-Sturges, D., Garcia, L., Lee, C., Zenick, H., Grevatt, P., Sanders, W. H., III, Case, H., & Dankwa-Mullan, I. (2011). Symposium on integrating the science of environmental justice into decision-making at the Environmental Protection Agency: An overview. *American Journal of Public Health, 101*(Suppl. 1), S19–S26. https://doi.org/10.2105/AJPH.2011.300368

Oesterreich, D. (2005). Flight into security: A new approach and measure of the authoritarian personality. *Political Psychology, 26*(2), 275–298. https://doi.org/10.1111/j.1467-9221.2005.00418.x

Olson-Buchanan, J. B., Drasgow, F., Moberg, P. J., Mead, A. D., Keenan, P. A., & Donovan, M. A. (1998). Interactive video assessment of conflict resolution skills. *Personnel Psychology, 51*(1), 1–24. https://doi.org/10.1111/j.1744-6570.1998.tb00714.x

Ones, D. S., & Dilchert, S. (2013). Measuring, understanding, and influencing employee green behavior. In A. H. Huffman & S. R. Klein (Eds.), *Green organizations: Driving change with I-O psychology* (pp. 115–148). Routledge.

Oskamp, S. (2000). A sustainable future for humanity? How can psychology help? *American Psychologist, 55*(5), 496–508. https://doi.org/10.1037/0003-066X.55.5.496

Patrick, V. M., & Hagtvedt, H. (2012). How to say "no": Conviction and identity attributions in persuasive refusal. *International Journal of Research in Marketing, 29*(4), 390–394. https://doi.org/10.1016/j.ijresmar.2012.04.004

Pergamin-Hight, L., Naim, R., Bakermans-Kranenburg, M. J., van IJzendoorn, M. H., & Bar-Haim, Y. (2015). Content specificity of attention bias to threat in anxiety disorders: A meta-analysis. *Clinical Psychology Review, 35*, 10–18. https://doi.org/10.1016/j.cpr.2014.10.005

Persons, J. B. (1993). The process of change in cognitive therapy: Schema change or acquisition of compensatory skills? *Cognitive Therapy and Research, 17*(2), 123–137. https://doi.org/10.1007/BF01172961

Peterson, B. E., & Zurbriggen, E. L. (2010). Gender, sexuality, and the authoritarian personality. *Journal of Personality, 78*(6), 1801–1826. https://doi.org/10.1111/j.1467-6494.2010.00670.x

Petkova, A. P., Rindova, V. P., & Gupta, A. K. (2013). No news is bad news: Sense-giving activities, media attention, and venture capital funding of new technology organizations. *Organization Science, 24*(3), 865–888. https://doi.org/10.1287/orsc.1120.0759

Pettigrew, T. F., & Tropp, L. R. (2008). How does intergroup contact reduce prejudice? Meta-analytic tests of three mediators. *European Journal of Social Psychology, 38*(6), 922–934. https://doi.org/10.1002/ejsp.504

Phillips, S. (2007, July 27). A brief history of Facebook. *The Guardian*. https://www.theguardian.com/technology/2007/jul/25/media.newmedia

Pike, G. R., Kuh, G. D., & Gonyea, R. M. (2007). Evaluating the rationale for affirmative action in college admissions: Direct and indirect relationships between campus diversity and gains in understanding diverse groups. *Journal of College Student Development, 48*(2), 166–182. https://doi.org/10.1353/csd.2007.0018

Pinchot, E. (2021). Learning with corporate sustainability leaders: Systemic barriers and collaborative openings to addressing climate change. *Dissertation Abstracts International: B. The Sciences and Engineering, 82*(1–B).

Poitras, J. (2012). Meta-analysis of the impact of the research setting on conflict studies. *International Journal of Conflict Management, 23*(2), 116–132. https://doi.org/10.1108/10444061211218249

Presotto, A., Verderane, M. P., Biondi, L., Mendonça-Furtado, O., Spagnoletti, N., Madden, M., & Izar, P. (2018). Intersection as key locations for bearded capuchin monkeys (*Sapajus libidinosus*) traveling within a route network. *Animal Cognition, 21*(3), 393–405. https://doi.org/10.1007/s10071-018-1176-0

Pulakos, E. D., Kantrowitz, T., & Schneider, B. (2019). What leads to organizational agility: It's not what you think. *Consulting Psychology Journal, 71*(4), 305–320. https://doi.org/10.1037/cpb0000150

Quelin, B. V., Cabral, S., Lazzarini, S., & Kivleniece, I. (2019). The private scope in public–private collaborations: An institutional and capability-based perspective. *Organization Science, 30*(4), 831–846. https://doi.org/10.1287/orsc.2018.1251

Rafaeli, A., & Sutton, R. I. (1987). Expression of emotion as part of the work role. *Academy of Management Review, 12*(1), 23–37. https://doi.org/10.5465/amr.1987.4306444

Raimi, K. T., Stern, P. C., & Maki, A. (2017). The promise and limitations of using analogies to improve decision-relevant understanding of climate change. *PLOS ONE, 12*(1), Article e0171130. https://doi.org/10.1371/journal.pone.0171130

Ramis, H. (Director). (1980). *Caddyshack* [Film]. Orion Pictures.

Rao, R. S., Chandy, R. K., & Prabhu, J. C. (2008). The fruits of legitimacy: Why some new ventures gain more from innovation than others. *Journal of Marketing, 72*(4), 58–75. https://doi.org/10.1509/jmkg.72.4.058

Rasmussen, H. N., Scheier, M. F., & Greenhouse, J. B. (2009). Optimism and physical health: A meta-analytic review. *Annals of Behavioral Medicine, 37*(3), 239–256. https://doi.org/10.1007/s12160-009-9111-x

Reese, G. (2012). When authoritarians protect the Earth—Authoritarian submission and proenvironmental beliefs: A pilot study in Germany. *Ecopsychology, 4*(3), 232–236. https://doi.org/10.1089/eco.2012.0035

Reid, T. R. (2010). *The healing of America*. Penguin.

Reser, J. P., & Swim, J. K. (2011). Adapting to and coping with the threat and impacts of climate change. *American Psychologist, 66*(4), 277–289. https://doi.org/10.1037/a0023412

Riemer, M., & Harré, N. (2017). Environmental degradation and sustainability: A community psychology perspective. In M. A. Bond, I. Serrano-Garcia, & C. B. Keys (Eds.), *APA handbook of community psychology: Vol. 2. Methods for community research and action for diverse groups and issues* (pp. 441–455). American Psychological Association. https://doi.org/10.1037/14954-026

Riemer, M., Voorhees, C., Dittmer, L., Alisat, S., Alam, N., Sayal, R., Bidisha, S. H., De Souza, A., Lynes, J., Metternich, A., Mugagga, F., & Schweizer-Ries, P. (2016). The Youth Leading Environmental Change project: A mixed-method longitudinal study across six countries. *Ecopsychology, 8*(3), 174–187. https://doi.org/10.1089/eco.2016.0025

Robert, C., & Wall, J. A. (2019). Humor in civil case mediations: A functional approach. *Humor, 32*(3), 361–391. https://doi.org/10.1515/humor-2017-0065

Roberts, A. (2017). *Tamed: Ten species that changed our world*. Hutchinson.

Rogelberg, S. G., & Gill, P. M. (2004). The growth of industrial and organizational psychology: Quick facts. *The Industrial-Organizational Psychologist, 42*, 25–27.

Rogelberg, S. G., O'Connor, M. S., & Sederburg, M. (2002). Using the stepladder technique to facilitate the performance of audioconferencing groups. *Journal of Applied Psychology, 87*(5), 994–1000. https://doi.org/10.1037/0021-9010.87.5.994

Rosen, B., Furst, S., & Blackburn, R. (2007). Overcoming barriers to knowledge sharing in virtual teams. *Organizational Dynamics, 36*(3), 259–273. https://doi.org/10.1016/j.orgdyn.2007.04.007

Rosling, H., Rosling, O., & Ronnlund, A. R. (2018). *Factfulness*. Flatiron.

Ross, A. (1924). *The fur hunters of the far west*. R.R. Donnelly Lakeside Press.

Rutjens, B. T., Sutton, R. M., & van der Lee, R. (2018). Not all skepticism is equal: Exploring the ideological antecedents of science acceptance and rejection. *Personality and Social Psychology Bulletin, 44*(3), 384–405. https://doi.org/10.1177/0146167217741314

Samba, C., Van Knippenberg, D., & Miller, C. C. (2018). The impact of strategic dissent on organizational outcomes: A meta-analytic integration. *Strategic Management Journal, 39*(2), 379–402. https://doi.org/10.1002/smj.2710

Sanchez-Burks, J., Karlesky, M. J., & Lee, F. (2015). Psychological bricolage: Integrating social identities to produce creative solutions. In C. E. Shalley, M. A. Hitt, & J. Zhou (Eds.), *The Oxford handbook of creativity, innovation, and entrepreneurship* (pp. 93–102). Oxford University Press.

Saxena, M. (2016, April 14–16). *IOers in the public interest* [Conference presentation]. Society for Industrial and Organizational Psychology 31st Annual Conference, Anaheim, CA, United States.

Schippers, M. C., Homan, A. C., & van Knippenberg, D. (2013). To reflect or not to reflect: Prior team performance as a boundary condition of the effects of reflexivity on learning and final team performance. *Journal of Organizational Behavior, 34*(1), 6–23. https://doi.org/10.1002/job.1784

Schommer, M., Richter, A., & Karna, A. (2019). Does the diversification–firm performance relationship change over time? A meta-analytical review. *Journal of Management Studies, 56*(1), 270–298. https://doi.org/10.1111/joms.12393

Schwabe, L., & Wolf, O. T. (2009). Stress prompts habit behavior in humans. *The Journal of Neuroscience, 29*(22), 7191–7198. https://doi.org/10.1523/JNEUROSCI.0979-09.2009

Schwabe, L., & Wolf, O. T. (2012). Stress modulates the engagement of multiple memory systems in classification learning. *The Journal of Neuroscience, 32*(32), 11042–11049. https://doi.org/10.1523/JNEUROSCI.1484-12.2012

Schwenk, C. R. (1990). Effects of devil's advocacy and dialectical inquiry on decision making: A meta-analysis. *Organizational Behavior and Human Decision Processes, 47*(1), 161–176. https://doi.org/10.1016/0749-5978(90)90051-A

Sedikides, C., & Gebauer, J. E. (2010). Religiosity as self-enhancement: A meta-analysis of the relation between socially desirable responding and religiosity. *Personality and Social Psychology Review, 14*(1), 17–36. https://doi.org/10.1177/1088868309351002

Sheeran, P., Harris, P. R., & Epton, T. (2014). Does heightening risk appraisals change people's intentions and behavior? A meta-analysis of experimental studies. *Psychological Bulletin, 140*(2), 511–543. https://doi.org/10.1037/a0033065

Singer, R. D., & Feshbach, S. (1959). Some relationships between manifest anxiety, authoritarian tendencies, and modes of reaction to frustration. *Journal of Abnormal and Social Psychology, 59*(3), 404–408. https://doi.org/10.1037/h0044511

Skitka, L. J., Hanson, B. E., & Wisneski, D. C. (2017). Utopian hopes or dystopian fears? Exploring the motivational underpinnings of moralized political engagement. *Personality and Social Psychology Bulletin, 43*(2), 177–190. https://doi.org/10.1177/0146167216678858

Smith, L. (1994). The Binet-Piaget connection: Have developmentalists missed the epistemological point? *Archives de Psychologie, 62*(243), 275–285.

Smith, L. G. E., Amiot, C. E., Smith, J. R., Callan, V. J., & Terry, D. J. (2013). The social validation and coping model of organizational identity development: A longitudinal test. *Journal of Management, 39*(7), 1952–1978. https://doi.org/10.1177/0149206313488212

Smither, J. W., London, M., & Reilly, R. R. (2005). Does performance improve following multisource feedback? A theoretical model, meta-analysis, and review of empirical findings. *Personnel Psychology, 58*(1), 33–66. https://doi.org/10.1111/j.1744-6570.2005.514_1.x

Snell-Rood, E. C., & Steck, M. K. (2019). Behaviour shapes environmental variation and selection on learning and plasticity: Review of mechanisms and implications. *Animal Behaviour, 147*, 147–156. https://doi.org/10.1016/j.anbehav.2018.08.007

Solberg Nes, L., Carlson, C. R., Crofford, L. J., de Leeuw, R., & Segerstrom, S. C. (2011). Individual differences and self-regulatory fatigue: Optimism, conscientiousness, and self-consciousness. *Personality and Individual Differences, 50*(4), 475–480. https://doi.org/10.1016/j.paid.2010.11.011

Songer, N., Kelcey, B., & Gotwals, A. (2009). How and when does complex reasoning occur? Empirically driven development of a learning progression focused on complex reasoning about biodiversity. *Journal of Research in Science Teaching, 46*(6), 610–631. https://doi.org/10.1002/tea.20313

Sosis, R. (2000). Religion and intragroup cooperation: Preliminary results of a comparative analysis of utopian communities. *Cross-Cultural Research, 34*(1), 70–87. https://doi.org/10.1177/106939710003400105

Star, S. A. (1949a). Problems of rotation and reconversion. In S. A. Stouffer & A. A. Lumsdaine (Eds.), *The American soldier* (Vol. II, pp. 456–519). Princeton University Press.

Star, S. A. (1949b). Psychoneurotic symptoms in the Army. In S. A. Stouffer & A. A. Lumsdaine (Eds.), *The American soldier* (Vol. II, pp. 411–455). Princeton University Press.

Steg, L., & Vlek, C. (2009). Encouraging pro-environmental behavior: An integrative review and research agenda. *Journal of Environmental Psychology, 29*(3), 309–317. https://doi.org/10.1016/j.jenvp.2008.10.004

Stouffer, S. A. (1949). Job assignment and job satisfaction. In S. A. Stouffer & A. A. Lumsdaine (Eds.), *The American soldier* (Vol. I, pp. 284–361). Princeton University Press.

Sturm, S. (2009). Negotiating workplace equality: A systemic approach. *Negotiation and Conflict Management Research, 2*(1), 92–106. https://doi.org/10.1111/j.1750-4716.2008.00030.x

Swim, J. K., Clayton, S., & Howard, G. S. (2011). Human behavioral contributions to climate change: Psychological and contextual drivers. *American Psychologist, 66*(4), 251–264. https://doi.org/10.1037/a0023472

Swim, J. K., Stern, P. C., Doherty, T. J., Clayton, S., Reser, J. P., Weber, E. U., Gifford, R., & Howard, G. S. (2011). Psychology's contributions to understanding and addressing global climate change. *American Psychologist, 66*(4), 241–250. https://doi.org/10.1037/a0023220

Szayna, T. S., Watts, S., O'Mahony, A., Frederick, B., & Kavanagh, J. (2017). *What are the trends in armed conflicts, and what do they mean for U.S. defense policy?* https://www.rand.org/pubs/research_reports/RR1904.html

Tang, M. (2019). Fostering creativity in intercultural and interdisciplinary teams: The VICTORY model. *Frontiers in Psychology, 10*, 2020. https://doi.org/10.3389/fpsyg.2019.02020

Tasca, G. A. (2021). Team cognition and reflective functioning: A review and search for synergy. *Group Dynamics, 25*(3), 258–270. https://doi.org/10.1037/gdn0000166

Tay, P. K. C., Jonason, P. K., Li, N. P., & Cheng, G. H.-L. (2019). Is memory enhanced by the context or survival threats? A quantitative and qualitative review on the survival processing paradigm. *Evolutionary Behavioral Sciences, 13*(1), 31–54. https://doi.org/10.1037/ebs0000138

Tedeschi, R. G. (1999). Violence transformed: Posttraumatic growth in survivors and their societies. *Aggression and Violent Behavior, 4*(3), 319–341. https://doi.org/10.1016/S1359-1789(98)00005-6

Terry, R. (2020, July 13). Travel is said to increase cultural understanding. Does it? *National Geographic.* https://www.nationalgeographic.com/travel/article/does-travel-really-lead-to-empathy

Thomas, L. A., & LaBar, K. S. (2008). Fear relevancy, strategy use, and probabilistic learning of cue-outcome associations. *Learning & Memory, 15*(10), 777–784. https://doi.org/10.1101/lm.1048808

Thomas, T., & Panchuk, M. (2009). Human rights campaign. In C. E. Stout (Ed.), *The new humanitarians: Inspiration, innovations, and blueprints for visionaries: Vol. 3. Changing sustainable development and social justice* (pp. 157–175). Praeger Publishers/Greenwood Publishing Group.

Thompson, C. T., Vidgen, A., & Roberts, N. P. (2018). Psychological interventions for post-traumatic stress disorder in refugees and asylum seekers: A systematic review and meta-analysis. *Clinical Psychology Review, 63*, 66–79. https://doi.org/10.1016/j.cpr.2018.06.006

Thornton, G. C., III, & Rupp, D. E. (2006). *Assessment centers in human resource management: Strategies for prediction, diagnosis, and development.* Erlbaum. https://doi.org/10.4324/9781410617170

Tjosvold, D. (2008). Constructive controversy for management education: Developing committed, open-minded researchers. *Academy of Management Learning & Education, 7*(1), 73–85. https://doi.org/10.5465/amle.2008.31413864

Tolman, E. C. (1948). Cognitive maps in rats and men. *Psychological Review, 55*(4), 189–208. https://doi.org/10.1037/h0061626

Tran, V., Garcia-Prieto, P., & Schneider, S. C. (2011). The role of social identity, appraisal, and emotion in determining responses to diversity management. *Human Relations, 64*(2), 161–176. https://doi.org/10.1177/0018726710377930

Tran, V., Páez, D., & Sánchez, F. (2012). Emotions and decision-making processes in management teams: A collective level analysis. *Revista de Psicología del Trabajo y de las Organizaciones, 28*(1), 15–24. https://doi.org/10.5093/tr2012a2

Triana, M. del C., Kim, K., Byun, S.-Y., Delgado, D. M., & Arthur, W., Jr. (2021). The relationship between team deep-level diversity and team performance: A meta-analysis of the main effect, moderators, and mediating mechanisms. *Journal of Management Studies, 58*(8), 2137–2179. https://doi.org/10.1111/joms.12670

Tsai, M. H., & Young, M. J. (2010). Anger, fear, and escalation of commitment. *Cognition and Emotion, 24*(6), 962–973. https://doi.org/10.1080/02699930903050631

Tsai, W., Chi, N., Grandey, A. A., & Fung, S. (2012). Positive group affective tone and team creativity: Negative group affective tone and team trust as boundary conditions. *Journal of Organizational Behavior, 33*(5), 638–656. https://doi.org/10.1002/job.775

Twenge, J. M., Campbell, W. K., & Foster, C. A. (2003). Parenthood and marital satisfaction: A meta-analytic review. *Journal of Marriage and the Family, 65*(3), 574–583. https://doi.org/10.1111/j.1741-3737.2003.00574.x

Umoh, R. (2017, November 9). *The crucial mindset Jeff Bezos says you should have if you want to be successful.* CNBC Make It. https://www.cnbc.com/2017/11/09/jeff-bezos-says-you-should-have-this-mindset-to-be-successful.html

U.S. Bureau of Labor Statistics. (2020). *Entrepreneurship and the U.S. economy.* https://www.bls.gov/bdm/entrepreneurship/entrepreneurship.htm

Van Leeuwen, M. H. D. (2016). *Mutual Insurance, 1550–2015.* Palgrave Macmillan. https://doi.org/10.1057/978-1-137-53110-0

Vicaria, I. M., & Dickens, L. (2016). Meta-analyses of the intra- and interpersonal outcomes of interpersonal coordination. *Journal of Nonverbal Behavior, 40*(4), 335–361. https://doi.org/10.1007/s10919-016-0238-8

Vroom, V. H., & Jago, A. G. (2007). The role of the situation in leadership. *American Psychologist, 62*(1), 17–24. https://doi.org/10.1037/0003-066X.62.1.17

Walker, L. J. (1971). *The discourses of Niccolo Machiavelli.* Penguin.

Wang, S., Liu, Y., & Shalley, C. E. (2018). Idiosyncratic deals and employee creativity: The mediating role of creative self-efficacy. *Human Resource Management, 57*(6), 1443–1453. https://doi.org/10.1002/hrm.21917

Weiss, H. M., & Cropanzano, R. (1996). Affective events theory: A theoretical discussion of the structure, causes and consequences of affective experiences at work. *Research in Organizational Behavior, 18,* 1–74.

Wenzel, E. (2021). Bill Gates wants you to step up on climate. *Greenbiz.* https://www.greenbiz.com/article/bill-gates-wants-you-step-climate

Wharton, A. S., & Erickson, R. J. (1993). Managing emotions on the job and at home: Understanding the consequences of multiple emotional roles. *Academy of Management Review, 18*(3), 457–486. https://doi.org/10.5465/amr.1993.9309035147

Wilke, G., & Thornton, C. (2019). Translucent boundaries, leaders, and consultants: How to work with whole organizations. In C. Thornton (Ed.), *The art and science of working together: Practising group analysis in teams and organizations* (pp. 224–236). Routledge/Taylor & Francis Group.

Wilson, D. S. (1998). Hunting, sharing, and multilevel selection: The tolerated-theft model revisited. *Current Anthropology, 39*(1), 73–97. https://doi.org/10.1086/204699

Woehler, M., Floyd, T. M., Shah, N., Marineau, J. E., Sung, W., Grosser, T. J., Fagan, J., & Labianca, G. (2021). Turnover during a corporate merger: How workplace network change influences staying. *Journal of Applied Psychology, 106*(12), 1939–1949. https://doi.org/10.1037/apl0000864

Wood, W., Lundgren, S., Ouellette, J. A., Busceme, S., & Blackstone, T. (1994). Minority influence: A meta-analytic review of social influence processes. *Psychological Bulletin, 115*(3), 323–345. https://doi.org/10.1037/0033-2909.115.3.323

Woodward, W. R. (1982). The 'discovery' of social behaviorism and social learning theory, 1870–1980. *American Psychologist, 37*(4), 396–410. https://doi.org/10.1037/0003-066X.37.4.396

Yang, Y. C., Boen, C., Gerken, K., Li, T., Schorpp, K., & Harris, K. M. (2016). Social relationships and physiological determinants of longevity across the human life span. *Proceedings of the National Academy of Sciences, 113*(3), 578–583. https://doi.org/10.1073/pnas.1511085112

Yong, K., Mannucci, P. V., & Lander, M. W. (2020). Fostering creativity across countries: The moderating effect of cultural bundles on creativity. *Organizational Behavior and Human Decision Processes, 157*, 1–45. https://doi.org/10.1016/j.obhdp.2019.12.004

Zajonc, R. B. (1984). On the primacy of affect. *American Psychologist, 39*(2), 117–123. https://doi.org/10.1037/0003-066X.39.2.117

Zare, M., & Flinchbaugh, C. (2019). Voice, creativity, and big five personality traits: A meta-analysis. *Human Performance, 32*(1), 30–51. https://doi.org/10.1080/08959285.2018.1550782

Zemel, A., Koschmann, T., & LeBaron, C. (2011). Pursuing a response: Prodding recognition and expertise within a surgical team. In J. Streeck, C. Goodwin, & C. LeBaron (Eds.), *Embodied interaction: Language and body in the material world* (pp. 227–242). Cambridge University Press.

Zemke, R. E. (1994). Training needs assessment: The broadening focus of a simple concept. In A. Howard (Ed.), *Diagnosis for organizational change: Methods and models* (pp. 139–151). Guilford Press.

Index

Abstracted thinking, 87
Accommodation, 91, 97
Active adaptation, 43
Adams, Douglas, 45
Adaptation(s), 23–41
   behavioral, 25–26
   and cognitive learning, 28–32
   and conflict, 72
   to conflict in social margins, 83–84
   deep, 33–39
   and ecotones, 106
   of environment, 21
   to evolutionary hacks, 79
   individual, 185–186
   and learning, 27–28
   managing resistance to, 134–135
   not reliant on genes, 8
   and persistence criterion of success, 151
   psychology and evolutionary theory, 26–27
   rate of adaptive change, 37, 39
   risk vs. opportunity in, 32–33
   and social nature of humans, 16–17
   social organizing as, 8–9
   and social speciation, 40–41
   supporting sustainable, 119–121
   when cultures meet, 101–102

Adaptive generalist social species, 191
American Psychologist, 158
Anxiety
   and authoritarianism, 136–137
   and creativity, 66
   and learning, 31
Appearances, managing, 137
Applied evolutionary psychology, 188
Applied psychology, 3–22
   approaches to sustainability based on, 12–14
   conflict management approaches in, 111
   criterion-centric approach, 14–16
   developmental change in, 85
   and ecotone management, 122, 124–125
   framework for sustainability from, 183
   framing of sustainability, 11–12
   management of social speciation process, 170–178
   processes in, 6–8
   research basis for practice, 19–20
   shared language in, 155–156
   social groups managed by, 59
   social organizing as adaptive, 8–9
   social organizing dynamics, 17–19

Applied psychology (*continued*)
  social organizing processes, 9–11
  and social speciation, 187–189
  social speciation in, 20–22
  sustainability as social problem, 16–17
Argyris, C., 58
Aristotle, 92
Arts, as communication medium, 114
Arvey, R. D., 64
Assessment centers, 87–89
Assessments
  in applied psychology processes, 6
  of behavioral change, and conformity, 139
  continuing importance of, 125
  of developmental change, 87–92, 127
  and feedback, 6, 12, 88, 132, 157, 159, 177
  of initial stakeholder perceptions of situation, 156–157
  and needs assessment, 12, 147
Assimilation, 91
Athens, city-state of, 148
Authoritarianism, 126–127, 136–138, 162, 164–165
"Authority = right" heuristic, 136
Automatic imitation, 102–103
Automatic thinking, 34

Barry, John, 60
Behavior(s)
  aspirations for, in conflict management, 131–133
  attributions of, 15–16
  changing, as objective of social speciation, 142
  changing individual, 190
  and conformity, 139
  control of, 10
  and implicit theories about change/conflict, 159–161
  learning, as adaptation, 49–50

learning to manage, 19
managing change of, 75, 76
and physical adaptations, 24–25
prosocial, 63–64
social hacks to moderate, 114–115
and social structures, 68–71
in support of social speciation processes, 181
and sustainability, 11–12
targeting, for social speciation, 125–126
targeting sustainability-related, 186–187
Behavioral adaptations, 25–26
Behavioral feedback, 132
Behavioral learning, 27
Behavioral practice, building self-efficacy with, 160
Behaviorism, 27–28
Bezos, Jeff, 93
Bias, 46–48
Big Five, 62
Binet, Alfred, 127
Biological adaptation, 24
Biological ecotones, 105
Blame, preoccupation with placing, 16
Boundary spanning leadership, 133–134
Brain. *See also* Hippocampus; Neocortex
  and evolutionary hacking, 38
  and learning, 25–26
  in research on learning, 28
Broad generalist species, 151
Burning Man, 118–119, 189

Carbon emissions, reducing, 188
Carrying capacity, 35–36
Carson, Rachel, 3, 8, 141, 191
Catastrophes, 12–14
Causal reasoning
  errors in, 45–48
  and risk, 33

INDEX

Cause and effect contingencies, 30
Caves, 78
Change
    behavioral, 75, 76, 139
    developmental. *See* Developmental change
    implicit theories about, 159–160
    normative, 140
Characters, 82
Cialdini, Robert, 34, 136, 139
Circumstances, in contingency rules, 48–49
Civil rights movement, 141
Cognitive development, stages of, 127
Cognitive learning, 28–32
Cognitive maps. *See* Mental maps
Cognitive models. *See* Mental maps
Colarelli, S. M., 64
Collective intelligence, 52
College experience, 92
Comfort, as motive, 185–186
Common physical spaces, 177
Communication. *See also* Language
    by other species, 67
    in social ecotones, 113–114
Commuter rail lines, 120
Competence, 121–123
Confirmation bias, 46–47, 69
Conflict
    and adaptation, 72
    and balance, 80
    among multiple social identities, 80
    defining, 128
    and developmental change, 71
    emotive aspects of, 128–130
    facilitating resolution of, 160
    implicit theories about, 160–161
    and mimicry in social ecotones, 103–105
    and prosperity, 110
    and science, 70–71
    and social expectations, 80–83
    and work–family, 80

Conflict management, 126–141
    authoritarian influence, 136–138
    behavioral aspirations, 131–133
    and boundary spanning leadership, 133–134
    and conformity, 138–139
    dealing with resistance, 134–135
    emotive aspects of conflict, 128–130
    by governments, 112–113
    and language, 113
    learning, from mimicry and television theater, 103
    and management of social speciation process, 155, 159
    normative minority influence, 139–141
    with perspective taking, 127–128
    recognition of expertise and, 130–131
    recognition of motives for, 156
    and scarcity, 110–111
    in social ecotones, 109–111, 114–116
Conformity, 138–139
Conformity heuristic, 139
Consolidation, of new social species, 174
Consumer psychology, 19
Contingency rules, 29, 45, 48–50
Conventions, 188–189
Cooling, control of, 5, 7–8, 77–79
Cooperation, encouraging, 128
Coping methods, 80
COVID-19 pandemic, 41, 153, 176
Creativity
    and anxiety, 66
    balancing, with conflict, 131
    at Google, 65
    and motives for individual adaptation, 185–186
    optimism and, 190–191
    as social, 64

223

INDEX

Criterion-centric approach, 5, 14–16, 147
    five categories in, 15–16, 19, 32, 181, 184
    five foes in, 15–16, 19, 32, 181, 184
Criterion dilemma, 5–6
Cultural learning, 52
Culture(s)
    and adaptation, 54, 101–102
    mimicry of other, 104–105
    mutual learning between, 108
Curiosity testing, 61–64, 76, 109, 160, 182

"Dancing bears," 48
Death, of social species, 148
Decision-making, about ethical issues, 165–166
Decision processes
    in applied psychology, 5
    shutdown of, 164
    steps in, 16–17
Deep acting, 82
Deep adaptation, 23, 33–39
Deliberative thinking, 34–35
Democracy, persistence of, 148
Developmental change, 84–97
    in adults, 90–92
    and authoritarianism, 126–127
    and conflict with others, 71
    difficulty assessing, 87–90
    interventions and, 76, 85, 97
    and mental maps, 97
    and normative minority influence, 140–141
    social components of, 94–97
    in social identities, 83
    and social speciation, 101
    and time perspective, 92–94
Developmental stages, 132–133, 139
Diamond, Jared, 72, 101, 113
Dinosaurs, 27
Discourses (Machiavelli), 137

Disposable plastic bags, 161
Diversification
    and conflict, 111, 125
    creating ecotones to increase, 177
    and social invention, 109
DNA change, 7, 25
Domestication, of animals, 35
Dormancy, for social species, 166
Double loop learning, 58
Downsizing, empowered teams approach to, 178–180

Eads, James, 60–61
Ecological laws, support of, 138
Ecotone management
    implicit theories about change in, 159–160
    shared language for, 155
    during transitions, 173
Education, developmental stage and, 127
Emotional labor, 82
"Emotivation," 128–129
Emotive aspects of conflict, 128–130
Emotive displays, 94–95, 129–130
Empowered teams, 178–180
Engineering, 6, 60–61, 68–69
Environment
    individual adaptation of, 185–186
    social species' attempt to alter, 65
Environmental concerns
    focusing on specific, 189–190
    lack of urgency surrounding, 12–13
Environmental pressures, modifications of, 7, 50–51, 65, 77, 78
Ethics
    in applied psychology research, 20
    and competence, 122
    at Facebook, 175
    and human exceptionalism, 44
    of managing transitions in new ecotones, 173

in research on adaptations, 25
and successful social speciation, 161–166
of using interventions, 119–121
utilitarian, 135
Evil social species, 162–165
Evolutionary biology, 135
Evolutionary hacking, 77–79
adapting to, 79
and human exceptionalism, 45–51
and social speciation, 21
Evolutionary theory
and psychology, 26–27
and reactive adaptation, 37
Expectations
assessing client, 147
contradictions of, 53–54
between social identities, 81, 84
Experts
recognition of, 130–131
reliance on, 116
Extinction, of social species, 148

Facebook, 65–66, 71, 175–176
Facial mimicry, 102
Facilitated conflict resolution, 160
Failure
defining, 147–148, 151
failed speciation with opportunity framing, 154
shared language to define, 156–157
Family, as safe places, 66
Feedback reports, 157
Fire, learning to control, 78
Flight
and analogy for human adaptive advantage, 24, 27, 67, 76, 119
development of, by humans, 27
evolution of, 24–25
forms to support processes for, 27–28, 181
success of flying machines, 149, 151

Floods, moderating effects of, 153
Florida, Richard, 109
Ford, Kevin, 96
Fossil fuels, reducing use of, 120
Framing, of sustainability, 11–12, 32–33
Freedom Riders, 141
Fulfillment, ethical questions related to, 161–162
Funder, David, 48
Future stakeholders, 125

Galileo Galilei, 70
Gandhi, Mohandas, 104
Gates, Bill, 176
Generalist social species
adaptive, 191
successful, 149–151
Geronimo, 104
Google, 65, 71
Government institutions, 112–113
Greece, ancient, 148
Group decision-making, 165–166
Group process
and feedback, 126, 157
and interventions, 76, 97, 123
leadership as, 16, 134, 139–140, 142
and shared mental models, 76, 97, 121
Guns, Germs, and Steel (Diamond), 72

Hakel, Milt, 87–89
Harari, Yuval Noah, 36, 62, 184
Harlem Cultural Festival, 118
Harrington, James, 154
Harvard University, 176
Heating, control of, 7–8, 77–79
Henrich, Joseph, 43, 52–54, 68
Heuristics, 32, 34, 45, 136, 139
Hippocampus, 26, 28
The Hitchhiker's Guide to the Galaxy (Adams), 45
Homo Deus (Harari), 36

Homo sapiens, importance of social speciation to survival of, 169
Hothersall, David, 70
Household resource conservation, 17
Human exceptionalism, 43–55
    and hacked adaptation, 45–51
    and social invention, 51–54
    and social speciation, 55
Humor, 53–54, 131, 179–180

Illinois Institute of Technology, 180
Imitation, automatic, 102–103
Implicit theories
    about change, 159–160
    about conflict, 160–161
Inclusion, as human social motive, 183
"Incubators," sustainable practice development at, 188
Individual adaptation, 81–83, 185–186
Individual behavior, changing, 190
Individual decision-making, 16
Individual developmental change, 71
Individual quality of life, 13–14, 51–52
Industrial and organization (I/O) psychology, 4
    conflict management approaches in, 111
    criterion-centric approach in, 147
    and ecotone management, 124
    growth of, 19
    minority influence of, 142–143
    social adaptation managed by, 57–58
    social speciation managed by, 142
    social species described by, 65
Influence (Cialdini), 34
"Inner doggie," 129
Innovation(s)
    and developmental change, 127
    and mental maps, 96–97
    and social organizing, 17–18
    speed of, 78–79
Insurance companies, 178
Intelligence, collective, 52

I/O psychology. *See* Industrial and organization (I/O) psychology
Isolated species, successful, 149

Kahneman, Daniel, 33, 129
King, Martin Luther, Jr., 141

Language
    development of, and social ecotones, 113–114
    required for social speciation, 67–68
    shared, 155–157
Latter-day Saints, 154
Lawrence of Arabia, 104
Leadership
    and assessment, 157
    boundary-spanning, 133–134
    as group process, 16, 134, 139–140, 142
    as minority influence, 139–141
    and shared mental models, 124, 125
Learning
    and automatic imitation, 102
    behavioral adaptation as, 25
    cognitive, 28–32
    developmental change vs., 85, 90
    and development of brains, 25–26
    and evolutionary theory, 26–27
    and feedback, 27, 88, 111, 132, 157
    as objective of social speciation, 142
    and RNA change, 25, 27, 186
    passed between social groups, 78
    safe places for, 63, 76–77
    in social groups, 59–60
    of social roles, 82–83
    and social support, 95–97
    through humor, 179–180
    various forms of, 27–28
Learning, as adaptive process, 7
Legal systems, assumptions of, 14
Legitimacy problem, for new ventures, 145–147

INDEX

Leonardo DaVinci, 149
LGBTQ+ behaviors, changing opinions about, 140
Liberal arts education, 174–175
Limbic systems, 129
Lollapalooza, 118
Long-term perspectives, 92–94
Loss aversion, 12

Machiavelli, Nicolo, 137
Magic, 53–54
Managing ourselves, 3, 8, 141
Manhattan Project, 108
Manipulation, 10
Mediation, shared language in, 155
Memory, stress and, 31–32
Mental health care, 18–19
Mental maps
   assessing changes in, 90
   and authoritarianism, 136
   changing, 75–76
   cognitive learning and, 28–30
   created by self-managing teams, 125
   and curiosity testing, 62–63
   and developmental change, 85–86, 97
   and innovations, 96–97
   limitations and advantages of, 40
   and mimicry, 102–103
   and neocortex, 45
   preentry preparation for, 172
   reducing error in, 48–50
   shared. *See* Shared mental models
   symbols related to, 72–73
Microsoft, 175–176
Migratory habits, 77–78
Milk, Harvey, 140
Mimicry
   and automatic imitation, 102–103
   and conflict in social ecotones, 103–105
   and conformity, 139
   in new businesses, 110–111

Minority groups, people from
   giving voice to, during speciation process, 173
   implicit beliefs about conflict for, 161
Minority influence, 139–143
Moderation of environmental pressures, 7–8, 9, 24, 40, 41, 65, 77, 79, 98–100, 137, 146, 152, 170, 185
   social pressures and, 110–112, 114–115, 146, 148–149
Monarchies, successful social speciation with, 162
Morris, Ian, 152
Mortality rate, for new social species, 145
Motivation(s)
   and cognitive learning, 31
   competing, in social margins, 84
   emotion and, 28–29
   optimism and, 190
   shared, and cooperation, 128
   shared motives for managing speciation process, 156–159
   in successful social speciation, 146, 152–154
   supporting motives criterion of success, 150–152
   understanding and acknowledging, 130
Motivational map, 156
Multiple stakeholder criterion problem, 146–147
   and referral, 6
Munificence, of ecotones, 111
Mutual benefit schemes, 178

National Institute of Mental Health, 18
National League of Cities Steering Committee for Energy, Environment, and Natural Resources, 4

Native Americans, 104
Needs, social species meeting, 100–101
Neocortex
    and learning, 25–26, 29–30
    and mental mapping, 45
    and social behavior, 49
New parents, experience of becoming, 90–91
New technology, withholding of, 165
Normative change, 140
Normative minority influence, 139–141

Oceana (Harrington), 154
"On the Primacy of Affect" (Zajonc), 129
Openness to experience (personality characteristic), 62
Opportunity, safety needed to seek, 32–33
Opportunity framing, 33, 154
Optimism, 190–191
Ostrom, Tom, 46–47

Parties, diverse people at, 114
Persistence criterion, of success, 148–149
Personality characteristics, 62
Perspective taking, 127–128
Physical form
    adaption of, 24–25
    new uses for, 50–51
Piaget, Jean, 91
Playfulness, 61–62, 65
Politics, threat framing in, 153–154
Positive transfer, 96
Preentry preparation, for social speciation process, 170–173
Privilege, resistance to change and, 135
Procedural memory systems, 32
Processes, in applied psychology, 6–8. *See also specific processes*
Proenvironmental behavior, 4

Profit
    business decisions to maximize, 165
    and mimicry, 110
Prosocial behaviors, 63–64
Prosperity, 109–110
Psychological interventions, 58, 76, 85, 111
    ethics of using, 119–121
    and feedback, 58, 74, 80, 111, 157
    and industrial and organizational psychology, 57–58
    to manage risk perception, 159–161
    for social speciation management, 170–172, 187
Psychological variables, in social speciation, 171, 172
Psychology
    and evolutionary theory, 26–27
    future of, 189–190
    growth in field of, 18–19
    reframing sustainability as issue in, 184–185
    as science, 26
Puns, 53
Purpose, redefining of, 162

Reactive adaptation, 36–37, 39, 43
Reality testing, 61–64
Reframing, for sustainability problem, 184–185
Religion(s)
    human exceptionalism and, 44
    "successful" social speciation in, 161–162
    thriving of, 149–150
Rendezvous, 104
Repatriation, 163
Research basis
    for psychological practice, 19–20
    for social speciation management, 119
Resistance, dealing with, 134–135
Resources, in biological ecotones, 106

Revolutions, social species arising from, 167
Rice agriculture, 180
Riemer, M., 141
Rising Tide (Barry), 60
Risk aversion
  and adaptation, 32–33
  and conflict, 134–135
  and development of social groups, 61
  and learning, 32
  and successful social speciation, 146
Risk perception
  interventions to manage, 159–161
  managing social speciation process via, 181–182
  manipulating, to motivate social speciation, 153–154
RNA change, 7, 25
Roberts, Alice, 35, 39, 184
Role definition, 160
Roosevelt, Theodore, 104–105

Safe places, 66–67
  conditions related to, 76–77
  and conformity, 139
  for learning, 63
Safety
  as human social motive, 183
  as motivation for individual adaptation, 185–186
  and motivation for successful social speciation, 153
  in social ecotones, and rate of social speciation, 186
Sapiens (Harari), 62
Saxena, Mahima, 180
Scarcity, 110–111
Schema, 91
Science
  languages used by, 67–68
  and social speciation, 68–69
  during World War II, 70–71

Scientist-practitioners
  ethics of using interventions, 119–121
  leadership roles for, in liberal arts education, 174–175
  management of social situations by, 76
  minority influence exerted by, 141
  social speciation managed by, 142
  taking on voice of future minority by, 173
Second-order social speciation, 79
Selection, as adaptive process, 7
Selection systems, 124
Self-efficacy
  for future, 190–191
  increasing, 160–161
Self-managing teams, 125–126
Sensory input, and adaptation, 29
Shared language, 155–157
Shared mental models, 52–54
  in applied psychology, 187
  as aspiration, 167, 174
  and boundary spanning leadership, 134
  and conflict, 76, 103, 124, 132–133, 135
  and development, 26, 97, 132–133
  development of, 7, 76, 176, 177
  and group process, 121, 132–133
  from individual adaptation, 185–186
  and learning, 34, 110
  and mimicry, 103
  and pivotal role in sustainability, 54, 76, 187
  in self-managing teams, 125
  for speciation within larger entities, 180
Shared motives, 156–159
Slavery, 163–165
Smartphones, 155
Social advantage, 31

Social distancing, 41
Social ecotones, 99–116. *See also*
    Ecotone management
  artificial, within larger entities,
    178–179
  automatic imitation and mimicry,
    102–103
  civil rights movement in, 141
  communication in, 113–114
  conflict and negotiation between,
    172
  conflict management in, 109–111,
    114–116
  contact and interaction between,
    171
  creation of, 112–113, 176–178,
    188–189
  defining, 99
  entry into, 171
  managing social speciation in, 170
  managing transitions in new, 173
  mimicry and conflict in, 103–105
  preparing for entry into, 171
  rate of social speciation and safety
    in, 186
  scarcity and conflict management,
    110–111
  self-managing teams as, 125–126
  successful isolated species in, 149
  understanding others' points of
    view in, 156
  as venues for speciation, 105–109,
    117
  and yarn bombers, 100–101
Social element, of cognitive learning,
  30–31
Social entities, social speciation
  within, 178–180
Social environment, curiosity testing
  in, 62–64
Social groups
  as adaptation, 72
  combining aspects of different, 101

  early emergence of, 57
  learning passed between, 78
  mimicry in, 103
  and stress, 66
Social ideas, development of,
  148–149
Social identities
  balancing the need for several,
    79–81
  and conformity, 139
  intentional management of, 119
  and social ecotones, 107
  and social margins, 79–81
  in social speciation, 58–61
Social invention
  and authoritarianism, 137
  and human exceptionalism,
    51–54
  new language for, 114
  rapid pace of recent, 122
Social learning, 49–50, 52
Social margins, 79–84
  adaptation to conflict, 83–84
  and developmental change, 97
  individual adaptation to, 81–83
  and social ecotones, 107
  and social identities, 79–81
Social motives, 158–159
Social organizing
  as adaptive, 8–9, 40
  and applied psychology, 6–8
  around mutually important
    problems, 108
  dynamics of, 17–19
  processes of, 9–11
  as social speciation, 10–11
  and sustainability initiatives, 17
  of youth, 141
Social problem, sustainability as,
  16–17
Social processes, as adaptive strategy,
  21, 38
Social roles, learning, 82–83

Social speciation, 57–74. *See also*
    Successful social speciation
  and adaptation, 40–41
  and applied psychology, 187–189
  in applied psychology, 20–22
  creating ecotones that support,
    177–178
  curiosity and reality testing in,
    61–64
  as deep adaptation, 24
  defining, 64–66
  describing process of, 71–73
  as evolutionary hacking, 37–38
  and human exceptionalism, 55
  ingredients for novel, 123
  language required for, 67–68
  and safe places, 66–67
  second-order, 79
  social ecotones as venues for,
    105–109
  social identities in, 58–61
  social organizing as, 10–11
  social structures and behaviors,
    68–71
  unmanaged, 174–176
Social speciation management,
    117–143
  competence in, 121–123
  and conflict management, 126–141
  and ecotone management, 122,
    124–125
  at steps in process, 170–178
  for successful outcome, 155–161
  supporting sustainable adaptation
    with, 119–121, 191
  targeting behaviors for, 125–126,
    186–187
Social speciation process, 169–182
  consolidation of new species, 174
  creating ecotones, 176–178
  describing, 71–73
  forms to support, 181
  within larger social entities, 178–180

  management of. *See* Social
    speciation management
  management of foes of
    sustainability, 181–182
  managing steps in, 170–178
  preentry preparation for, 170–173
  transitions in new ecotones, 173
  unmanaged speciation in,
    174–176
Social species
  creation of, 172
  defining, 64–65
  engineering as, 60–61
  existing motive support by,
    150–152
  extinction of, 148
  persistence of, 148–149
  supporting consolidation of, 174
  survival of, 59–60, 151
  thriving by, 149–150
  yarn bombers as, 100–101
Social structures
  adaptation to, 7
  as adaptive strategy, 21, 38
  and behaviors, 68–71
  and causal reasoning, 47
  changes in, related to COVID-19
    pandemic, 41
  clashes of, 69–71
  curiosity testing reliant on, 63
  to manage conflict, 115
  and physical inventions, 72
  as reactive vs. deliberate, 41
  for science, 69
  social identity in, 58–59
  species that can alter, 40–41
  testing social invention in, 51–52
Social support, 76–77
  in conflict management, 129–130
  for consolidation of new species,
    174
  and developmental change, 95–97
  learning and, 49–50, 52

Social support (*continued*)
　managing speciation process with, 155
　and new social species, 101
　social ecotones created for, 113
　for social speciation process, 181
　for successful social speciation, 154
Social system, speciation within, 178–180
Socrates, 70
Stakeholders
　assessing initial perceptions of, 156–157
　considering future, 125
　determining relevant, 10
　future, projecting voice of, 173
　and successful social speciation, 146–147
Start-ups, social speciation in, 175–176
Stimulus–response learning, 27–28
Stonewall Union, 140
Stress
　and creativity, 66
　and environmental pressures, 77
　learning and, 31–32
　as motivation for successful social speciation, 153–154
Subrules, for mental maps, 48–50
Success
　criteria for, 147–151
　shared language for defining, 156–157
Successful social speciation, 145–167
　ethical questions about, 161–166
　failure of, 148
　managing process for, 155–161
　motivations for, 152–154
　persistence criterion for, 148–149
　supporting motives criterion for, 150–152
　thriving criterion for, 149–150
Supporting motives criterion, of success, 150–151

Surprise, management of, 115
Survival, of social species, 151
Sustainability, 183–191
　applied psychology and social speciation, 187–189
　approaches to, 12–14
　behavioral causes of, 11–12
　cognitive learning related to, 31–32
　definitions of, 5
　five categories for, 15–16, 19, 32, 181, 184
　five foes of, 15–16, 19, 32, 181, 184
　framing of, 11–12
　and future of psychology, 189–190
　and individual adaptation, 185–186
　and individual quality of life, 13–14
　lack of solutions for, 13
　management of foes of, 181–182
　optimism and self-efficacy for future, 190–191
　reframing problem of, 184–185
　research on behaviors related to, 4–5
　and shared mental models, 167
　of social motives, 158–159
　social organizing for, 8–9
　as social problem, 16–17
　targeting behaviors related to, 186–187
　thriving of species organized around, 150–152
Symbols, 73
Systems engineering, 27–28

Tamed (Roberts), 35
Telephone technologies, 165
Television theater, 103
Temperature control, 7–8, 77–79
Theory, working with, 20
Thinking
　abstracted, 87
　control of, 10
　deliberative and automatic, 34
　and social speciation, 61

INDEX

Thorndike, Edward, 28
Threat framing, 152–154
Thriving, as criterion of success, 149–150
Time, as measure of successful social speciation, 148–149, 151
Time perspective, 92–94
Tjosvold, Dean, 128, 131
Tolman, Edward, 28–29
Transfer of training, 96
Transitions, in new ecotones, 173
Travel, developmental change and, 93

Uncertainty, about social standing, 138–139
Underground Railroad, 164
Unethical tactics, 165–166
Unmanaged speciation, 174–176
Unmet needs
   developing organizations to address, 152
   exacerbation/maintenance of, 161–162

Unregulated social speciation, 169
Urban farms, 174
Urgency, of environmental concerns, 12
Utilitarian ethics, 135
Utopian ideals, 154

Venture capital firms, evaluating viability of new firms for, 188
Virtual teams, 122, 124

Wake-up call, stakeholder assessments as, 157
Western Europeans, 104
Woodstock, 118
Work–family balance, 80
Wright brothers, 151

X, Malcolm, 140

Yarn bombers, 100–101, 115, 177

Zajonc, Robert, 129
Zuckerberg, Mark, 176

About the Author

Robert G. "Bob" Jones, PhD, is Emeritus Professor at Missouri State University. He earned a bachelor of arts degree from St. Olaf Paracollege, concentrating on the clashing political theories that defined major events in Tudor and Stuart England. After graduation, he worked for the Chicago Children's Choir and pursued a career in music, with "straight jobs" in banking. After moving to Columbus, Ohio, he completed master's (1989) and doctoral (1992) degrees in industrial and organizational psychology from The Ohio State University.

Coming from a prominent family of scientists, jurists, and politicians on one side and musicians, public servants, and traditional outsiders on the other, Bob learned early that he would never solve the world's problems but that he could lend his weight to efforts to move things forward. This is what he has tried to do, starting as a musician in Chicago during civil rights efforts from the 1960s into the 1980s and then switching to become an applied psychologist, teacher, consultant, and public servant from the 1990s to the present.

As both a scholar and practitioner, he has wrestled with problems of managing natural human systems of exclusion, emotive perception and responding, family preference, and—for the past decade—the human inclination toward social invention, which has led us to sacrifice our collective future in the interest of current individual comfort. Bob has taught, presented, consulted, and published numerous books and articles, mostly relating to these natural system management problems, in the hope that readers will find help to push toward a better future.

Bob has served in various consulting capacities in the insurance, retail, transportation, publishing, and not-for-profit sectors. He has also served as a Springfield, Missouri, city council member and in numerous other volunteer roles.

Bob's wife is Professor Emeritus from Missouri State University, too, and their two sons are engaged in careers related to education—one as a high school history teacher in Washington, DC, and the other as a film and TV writer and producer. Bob and his wife are avid hikers in his home state of Montana, and he still does music—most recently releasing an album of original songs (*For the Duration* on the Guilt by Association label, available on most music streaming services).